T0406690

Marketing Food Brands

Ranga Chimhundu

Marketing Food Brands

Private Label versus Manufacturer Brands in the Consumer Goods Industry

Ranga Chimhundu
School of Management & Enterprise
University of Southern Queensland
Toowoomba, QLD, Australia

ISBN 978-3-319-75831-2 ISBN 978-3-319-75832-9 (eBook)
https://doi.org/10.1007/978-3-319-75832-9

Library of Congress Control Number: 2018939413

Printed on acid-free paper

This Palgrave Macmillan imprint is published by the registered company Springer International Publishing AG part of Springer Nature.
The registered company address is: Gewerbestrasse 11, 6330 Cham, Switzerland

Acknowledgements

Writing a book of this nature is a huge undertaking that puts demands not only on the author but on other people as well. I am indebted to many individuals and organisations for their contributions of varying degrees. I would therefore like to express my sincere gratitude to everyone who assisted in one way or another and whose contributions and sacrifices helped to enable its completion. While it would not be possible to mention all by name, and the following list is by no means exhaustive, special mention must go to the following: Dr Rob Hamlin and Associate Professor Lisa McNeill, for their advice on the large research project that forms the basis of this book. Research participants from the New Zealand FMCG industry and their respective organisations, as well as other participants who supplied data. The University of Otago, School of Business, for the scholarship award. Professor Mark Durkin and Professor Brendan Gray, discussants at a research colloquium during the very early stages of the research. Professor Jill Sweeney, for comments on the research at a workshop on publishing. Associate Professor Ron Garland and Associate Professor Lorraine Friend, as well as other participants at a conference where part of the material for the book was presented and discussed. The editors and anonymous reviewers of the *Journal of Brand Management*, *Australasian Marketing Journal*, *British Food Journal*, *Asia Pacific Journal of Marketing and Logistics*, *Journal of Euromarketing*, *International Journal of Business and Management* and *International Journal of Marketing Studies*, where papers based on the larger project that forms this book were accepted and published. The scrutiny of the manuscripts submitted and the necessary revisions thereafter enabled the refinement of this book not only from a central argument/storyline point of view, but also from the viewpoint of its incremental contribution to the current state of knowledge in the research area, over and above the published articles. Further thanks go to the University of Southern Queensland, for creating a research environment that encourages, enables and supports academic researchers to publish in high-quality research publications. Thanks to my family, for all their support. Finally, I wish to express my gratitude to all individuals who gave words of encouragement.

Contents

List of Abbreviations

BCG	Boston Consulting Group
CC	category captain
CEO	chief executive officer
ECR	efficient consumer response
FGC	Food and Grocery Council
FMCG	fast-moving consumer goods
IGD	Institute of Grocery Distribution
NPD	new product development
OECD	Organisation for Economic Co-operation and Development
R&D	research and development
RPM	resale price maintenance
SKU	stock-keeping unit

List of Figures

List of Tables

1

Introduction to Issues around Marketing Private Label and Manufacturer Brands

Overview

This chapter covers introductory issues related to research on the coexistence of private label and manufacturer brands in food product categories; these categories are also referred to as fast-moving consumer goods (FMCG) categories. The playing field for both private label and manufacturer brands is supermarket shelves. The research investigates how manufacturer brands and private label coexist in FMCG/supermarket product categories in a grocery retail landscape characterised by high retail concentration, and how relevant power is to this coexistence. Power, in this regard, is the influence or control that the two types of brands have on each other. Specifically, the chapter summarises the purpose and background of this research, its significance, the methodology employed in the investigation, the scope of the book and the contribution it makes.

R. Chimhundu, *Marketing Food Brands*,
https://doi.org/10.1007/978-3-319-75832-9_1

Purpose and Background of this Book

The purpose of this book is to investigate the coexistence of private label and manufacturer food brands in FMCG product categories in a grocery retail landscape characterised by high retail consolidation and concentration, and direct competition between brands owned and managed by owners of the grocery retail shelves (private label) and those owned and managed by their suppliers (manufacturer brands). The literature reviewed in Chaps. 2, 3 and 4 of this book leads to the development of research questions and related issues. The primary research question of the investigation is: *How do manufacturer brands and private label coexist in FMCG/supermarket product categories in a grocery retail landscape characterised by high retail concentration, and how relevant is power to this coexistence?*

This primary research question is broken down into three subsidiary research questions:

- Does a grocery retail environment characterised by high retail concentration lead to an overdominance of private label in relation to manufacturer brands in FMCG/supermarket product categories?
- How important are aspects of strategic dependency between manufacturer brands and private label in determining the nature of coexistence between the two types of brands in FMCG/supermarket product categories?
- In an FMCG/supermarket landscape characterised by high retail concentration and direct competition between brands owned and managed by owners of the grocery retail shelves (private label) and those owned and managed by their suppliers (manufacturer brands), what role is played by power in the coexistence relationship between the two types of brands in the product categories?

It is important to understand these issues because private label brands have become a common feature in FMCG/supermarket categories, and have also become a global phenomenon (ACNielsen 2005; Nielsen 2014). There is an increasing prominence of private label brands in supermarket product categories. Private label brands, however, belong to the grocery retailers, who own the retail shelves and are customers of the manufacturers; but at the same time these brands are competing with manufacturer brands that are owned by the grocery retailers' suppliers (i.e. FMCG manufacturers). How does this coexistence work? This situation most certainly implies intricate relationships in the coexistence of manufacturer brands and private label brands, both of which have become a permanent feature in most supermarket categories around the world. It is noted that in the FMCG manufacturer–retailer relationship, the balance of power has largely shifted to the retail chains because of a number of factors that include retail consolidation and concentration, the power of information on the part of the retailers, and the retailers' increasing emphasis on private label (e.g. Hogarth-Scott 1999; Hollingsworth 2004; Hovhannisyan and Bozic 2016; Stanković and Končar 2014; Sutton-Brady et al. 2015; Weitz and Wang 2004). Retail consolidation and concentration have been linked with driving private label brand shares to high levels (e.g. Burt 2000; Cotterill 1997; Defra 2006; Hollingsworth 2004; Nielsen 2014; Rizkallah and Miller 2015). If the grocery retail chains have the power and will, they may decide to go all the way in terms of private label dominance. Some researchers (Hoch et al. 2002a, b) have noted that the private label is privileged in the way it controls its own marketing mix as well as the marketing activities of competitor brands. This situation has made it necessary to conduct research that seeks to better understand private label and manufacturer brand share trends (Chimhundu et al. 2011), given that the odds seem to be stacked against manufacturer brands in what one study (Kumar and Steenkamp 2007) has referred to as the radically altered FMCG landscape.

Additionally, an intensive investigation of the coexistence of the two types of brands in selected food product categories was seen as a good way to establish a richer picture of the situation, including an assessment of what one author (Baden-Fuller 1984: 525) has referred to as "retail strategic thinking in the area of retail brands". Furthermore, in the environment of retailer power, consumer focus is stressed in the category management literature as a factor that is commonly recognised as having an influence on the balancing of manufacturer brands and private label in the product categories (e.g. ACNielsen et al. 2006). The mainstream academic literature has, however, fallen short of giving adequate attention to other key factors such as the influence of aspects of strategic dependency between the two types of brands on the determination of how they coexist. In view of the conflicting views in the literature on the relative contributions of manufacturer brands and private label to product innovation in the categories (Aribarg et al. 2014; Coelho do Vale and

Verga-Matos 2015; Conn 2005; Hoch and Banerji 1993; Lindsay 2004; Olbrich et al. 2016; Silverman 2004; Steiner 2004) and category support (e.g. Anonymous 2005; Putsis and Dhar 1996), this book further investigates these issues with a view to arguing that aspects of strategic dependency that enhance private label in the categories have relevance for the determination of policies that govern the coexistence of the two types of brands in FMCG/supermarket categories. This research direction, having been initially refined and reported by Chimhundu et al. (2010), has now undergone full development in this volume.

The explicit theory employed in this research is that of power (Dapiran and Hogarth-Scott 2003; French and Raven 1959; Hunt 2015; Mintzberg 1983; Raven 1993). This is in view of the power wielded by the retail chains in relation to manufacturers, as has already been mentioned. The bases of power are examined in relation to the coexistence of manufacturer brands and private label, and particularly in view of the radically altered (Kumar and Steenkamp 2007) grocery retail environment. The research seeks to establish the dominant source(s) of power employed, as well as the implications for the continued coexistence of the two types of brands in the categories.

Significance of the Book

The context of this research book is the New Zealand FMCG environment, but the lessons derived from the book have global implications. According the New Zealand Food and Grocery Council (FGC), an industry association that represents the manufacturers and suppliers of New Zealand's food, beverage and grocery brands, the FMCG sector plays a major role in the New Zealand economy.

> FGC members "represent more than NZ$34 billion in domestic retail sales, more than NZ$31 billion in exports, and directly or indirectly employ about 400,000 people, or one in five people in our workforce. The NZ$31 billion in exports was 72 per cent of New Zealand's total merchandise exports in 2014 to 195 countries. Food and beverage manufacturing is the largest manufacturing sector in New Zealand, representing 44% of total manufacturing income". (http://www.fgc.org.nz/: accessed 31 May 2016)

This book therefore addresses a major industry that contributes significantly to the New Zealand economy; furthermore, on a global scale the industry is also vast and has great economic significance.

In addition, the private label industry has become a huge industry globally. Private label brands have been estimated as having an approximately 17% share of the global market (Nielsen 2014). Grocery retailers with strong private label products across many categories have continued to challenge manufacturer brands in the minds of consumers (ACNielsen 2005).

Similarly, the size and importance of the private label industry can be further illustrated by examining a specific market/country that has a high food and grocery private label share by world standards: the United Kingdom (UK). The UK has approximately 41% private label share (Nielsen 2014). In the UK grocery industry, which is worth approximately £192.6 billion (statistica.com, accessed 9 December 2017), private label can be estimated at £79 billion, which is phenomenal. Furthermore, the Institute of Grocery Distribution (IGD) estimated that consumers in Europe were going to be spending approximately €430 billion on private label brands in 2010 (IGD 2006) and that figure has gone much higher in 2018. All these facts serve to demonstrate the arrival and importance of private label around the world alongside manufacturer brands.

In many countries, FMCG private label brands have become an established feature of supermarket shelves and some retail chains have stepped up their marketing effort behind private label. In the country of this study, New Zealand, all major supermarket chains have private label programmes in place in many food and grocery product categories. In this regard, this in-depth study of the coexistence of manufacturer brands and private label addresses an important area, and advances knowledge in the research direction outlined in the preceding section on the purpose and background of the research. Further discussion of its research contribution comes later in this chapter.

The lessons learnt from this study have worldwide implications, with the pattern of such implications varying depending on how the characteristics of the industry (or country) in question resemble or differ from the characteristics of the country of this intensive study. Both manufacturer brands and private label are active participants in this industry. Academic research that seeks to better understand specific aspects of this industry in a way that advances new knowledge is worth the effort given the industry's economic significance.

Methodology Employed in the Investigation

With regard to the research paradigm, the author considered the interpretive paradigm (and its three variants: critical theory, constructivism and realism) and the positivist paradigm, and chose to adopt the interpretive paradigm (realist variant). The book therefore operates largely from an interpretive, realist perspective. Realism was chosen as the most appropriate paradigm for the book as it enables the exploration of both observable and non-observable phenomena and, in addition, it makes room for the incorporation of a variety of sources of evidence in addressing the "whats", "hows" and "whys" of the coexistence of manufacturer brands and private label in food product categories.

The appropriate methodology was judged to be the case study research methodology (Yin 2003, 2013). This methodology was chosen for a number of reasons, including its ability to tackle "how" and "why" questions (Robson 1993; Yin 2003, 2013), and its capacity to allow the researcher to acquire deep and detailed qualitative data by getting closer to the phenomenon. Furthermore, the case research method can "draw on a wider array of documentary information" in addition to the employment of interviews (Yin 2003: 6) to address research issues.

The research made use of a preliminary study into private label share trends in four countries, followed by a pilot study, and then the main study which was New Zealand based. The main study involved the collection of data from food retailers, FMCG manufacturers and consultants. Data collection was done through a combination of research interviews, in-store category observation and the location and reviewing of relevant documentation. The data were subjected to content analysis and the research results and discussion are structured along the lines of the research issues developed from the literature.

Scope of the Book

The preliminary study focused on aggregate private label share trends in four developed economies; New Zealand, Australia, the UK and the USA, from 1992 to 2005. The pilot study, which was conducted in preparation for the main study, was largely New Zealand based and consisted of five interviews (two with food retailers, one with a manufacturing company, and two with industry analysts), as well as two category observation exercises in supermarkets. The main study focused on the New Zealand FMCG/supermarket industry and incorporated retail chains, FMCG manufacturers and consultants. It examined five food categories that included private label brands: namely, milk, flour, cheese, breakfast cereals and tomato sauce. In all, 46 in-depth interviews were held (30 with FMCG retailers, 13 with FMCG manufacturers/suppliers, and three with consultants); further, in-store category observation exercises were carried out in each of the respective product categories in 18 stores. With respect to the drawing of conclusions from the book, while the scope of the research is limited to food product categories, grocery retail chains and the FMCG/supermarket industry of one economy, and generalisability is analytic rather than statistical because of the nature of the study, the findings of the book and the lessons learnt are relevant to the global audience, and specifically to all countries that have both manufacturer brands and private label on food and grocery retail shelves.

Contribution of the Book

This book makes a contribution to knowledge in the area of the coexistence of manufacturer brands and private label in a number of related ways. Firstly, it is established that an environment characterised by high retail consolidation and concentration does not necessarily lead to an overdominance of private label (over manufacturer brands) in FMCG/supermarket product categories (Chimhundu et al. 2011). There are equilibrium points in the categories beyond which grocery retailers may not want to go with their private label, and the points serve to safeguard the long-term strategic health of the categories.

Secondly, while consumer choice is prominent in the academic literature (through its central role in category management) as a factor that plays a part in determining the nature of coexistence between manufacturer brands and private label, this study illustrates that other deeper underlying factors that have not been given much prominence in the mainstream academic literature are at play as well. Aspects of strategic dependency between manufacturer brands and private label have relevance for the determination of policies that govern the coexistence of the two types of brands in FMCG/supermarket product categories. These aspects are the respective brand types' comparative capacity to deliver product innovation in the categories, and their comparative capacity to deliver category support (category development) in the categories. Manufacturer brands' greater collective capacity to deliver product innovation and category support is a key aspect of strategic dependency that shapes the nature of coexistence between the two types of brands (Chimhundu et al. 2015).

Thirdly, in an FMCG/supermarket landscape characterised by high retail consolidation/concentration and direct competition between private label and manufacturer brands, the theory of power is intricately connected to the coexistence of the two types of brands. Despite the balance of power being largely in favour of grocery retailers, it is typically expert and referent bases of power rather than coercive power that are dominant in the coexistence relationship of the two types of brands (Chimhundu 2016). The book further offers an integrated conceptual framework, based on empirical evidence, which suggests a number of research propositions for further development into hypotheses that can be investigated via confirmatory studies set in the positivist paradigm.

These research contributions offer "smaller bricks of new knowledge" (Lindgreen 2001: 513), and each of them takes an "incremental step in understanding" (Phillips and Pugh 2000: 64) the coexistence of manufacturer brands and private label in FMCG/supermarket product categories.

Book Outline

Other than the preliminaries and the reference matter, the text component of this book is organised as follows.

Chapter 1

This chapter has served to introduce the book by giving an overall picture of its subject matter. Briefly discussed are the main aspects of the book, including its purpose and background, significance, methodology, scope and contribution.

Chapters 2, 3 and 4

These chapters review the literature on the research topic within parent and related disciplines. The range of literature covers category management; innovation, category support and consumer choice; power; and manufacturer and retailer brands in FMCG/supermarket product categories. In the process, the study's primary research question and subsidiary research questions as well as specific research issues are developed.

Chapter 5

This chapter further develops the research direction created with respect to research issues. The issues have three main themes: the balance between manufacturer brands and private label in the food product categories; innovation, category support and consumer choice; and category strategic policies governing the coexistence of manufacturer brands and private label in the product categories. In addition, a graphical conceptual framework is created.

Chapters 6 and 7

These chapters discuss the research process followed in addressing the research issues developed for this book. In this respect, the research paradigm and research approach are chosen and justified. Furthermore, the study's research design and execution are discussed.

Chapter 8

This chapter discusses the results of the main study on the coexistence of manufacturer brands and private label in FMCG/supermarket product categories. The data collected through interviews, in-store category observation and documentation are analysed and discussed against the research issues in Chap. 5. In addition, some aspects of the preliminary study on private label brand share trends are integrated into addressing some of the research issues.

Chapter 9

This concluding chapter discusses a number of aspects, including the findings of the book in the context of the literature, the resultant conceptual framework, the theoretical implications and implications for marketing practice, and the conclusions that may be drawn from the book, as well as directions for further research and the key contributions of the book.

Chapter Recap

This chapter has covered the purpose and background of this book, as well as describing its significance, the methodology employed in the research that forms its basis, its scope, its contribution and its outline. The next chapter looks at the management of FMCG product categories.

References

ACNielsen. (2005). *The power of private label: A review of growth trends around the world*. New York, NY: ACNielsen.

ACNielsen, Karolefski, J., & Heller, A. (2006). *Consumer-centric category management: How to increase profits by managing categories based on consumer needs*. Hoboken, NJ: Wiley.

Anonymous. (2005). House brand strategy doesn't quite check out. *The Age*. Retrieved December 3, 2017, from http://www.theage.com.au/news/Business/House-brand-strategy-doesnt-quite-check-out/2005/04/01/1112302232004.html

Aribarg, A., Arora, N., Henderson, T., & Kim, Y. (2014). Private label imitation of a national brand: Implications for consumer choice and law. *Journal of Marketing Research, 51*(6), 657–675.

Baden-Fuller, C. W. F. (1984). The changing market share of retail brands in the UK grocery trade 1960–1980. In C. W. F. Baden-Fuller (Ed.), *The economics of distribution* (pp. 513–526). Milan: Franco Angeli.

Burt, S. (2000). The strategic role of retail brands in British grocery retailing. *European Journal of Marketing, 34*(8), 875–890.

Chimhundu, R. (2016). Marketing store brands and manufacturer brands: Role of referent and expert power in merchandising decisions. *Journal of Brand Management, 23*(5), 24–40.

Chimhundu, R., Hamlin, R. P., & McNeill, L. (2010). Impact of manufacturer brand innovation on retailer brands. *International Journal of Business and Management, 5*(9), 10–18.

Chimhundu, R., Hamlin, R. P., & McNeill, L. (2011). Retailer brand share statistics in four developed economies from 1992 to 2005: Some observations and implications. *British Food Journal, 113*(3), 391–403.

Chimhundu, R., McNeill, L. S., & Hamlin, R. P. (2015). Manufacturer and retailer brands: Is strategic coexistence the norm? *Australasian Marketing Journal, 23*(1), 49–60.

Coelho do Vale, R., & Verga-Matos, P. (2015). The impact of copycat packaging strategies on the adoption of private labels. *Journal of Product & Brand Management, 24*(6), 646–659.

Conn, C. (2005). Innovation in private label branding. *Design Management Review, 16*(2), 55–62.

Cotterill, R. W. (1997). The food distribution system of the future: Convergence towards the US or UK model? *Agribusiness, 3*(2), 123–135.

Dapiran, G. P., & Hogarth-Scott, S. (2003). Are co-operation and trust being confused with power? An analysis of food retailing in Australia and New Zealand. *International Journal of Retail and Distribution Management, 31*(5), 256–267.

Defra (Department for Environment, Food and Rural Affairs). (2006). *Economic note on UK grocery retailing*. London, UK: Department for Environment, Food and Rural Affairs, Food and Drinks Economics Branch.

French, J. R. P., & Raven, B. (1959). The bases of social power. In D. Cartwright (Ed.), *Studies in social power*. Institute of Social Research (pp. 150–167). Ann Arbor, MI: The University of Michigan.

Hoch, S. J., & Banerji, S. (1993). When do private labels succeed? *Sloan Management Review, 34*(4), 57–67.

Hoch, S. J., Montgomery, A. L., & Park, Y. H. (2002a). Why private labels show long-term market evolution. *Marketing Department Working Paper*, Wharton School, University of Pennsylvania, PA.

Hoch, S. J., Montgomery, A. L., & Park, Y. H. (2002b). Long-term growth trends in private label market shares. *Marketing Department Working Paper #00-010*, Wharton School, University of Pennsylvania, PA.

Hogarth-Scott, S. (1999). Retailer-supplier partnerships: Hostages to fortune or the way forward for the millennium? *British Food Journal, 101*(9), 668–682.

Hollingsworth, A. (2004). Increasing retail concentration: Evidence from the UK food sector. *British Food Journal, 106*(8), 629–638.

Hovhannisyan, V., & Bozic, M. (2016). The effects of retail concentration on retail dairy product prices in the United States. *Journal of Dairy Science, 99*(6), 4928–4938.

Hunt, S. D. (2015). The bases of power approach to channel relationships: Has marketing's scholarship been misguided? *Journal of Marketing Management, 31*(7–8), 747–764.

IGD (Institute of Grocery Distribution). (2006). Own labels will be worth €430bn by 2010. Retrieved December 4, 2017, from https://www.igd.com/

Kumar, N., & Steenkamp, J. E. M. (2007). *Private label strategy: How to meet the store brand challenge*. Boston, MA: Harvard Business School Press.

Lindgreen, A. (2001). A framework for studying relationship marketing dyads. *Qualitative Market Research: An International Journal, 4*(2), 75–87.

Lindsay, M. (2004). Editorial: Achieving profitable growth through more effective new product launches. *Journal of Brand Management, 12*(1), 4–10.

Mintzberg, H. (1983). *Power in and around organisations*. Englewood Cliffs, NJ: Prentice Hall.

Nielsen. (2014). *The state of private label around the world*. The Nielsen Company.

Olbrich, R., Hundt, M., & Jansen, H. C. (2016). Proliferation of private labels in food retailing: A literature overview. *International Journal of Marketing Studies, 8*(8), 63–76.

Phillips, E. M., & Pugh, D. S. (2000). *How to get a PhD: Handbook for students and their supervisors* (3rd ed.). Buckingham: Open University Press.

Putsis, W. P., & Dhar, R. (1996). Category expenditure and promotion: Can private labels expand the pie? *Working Paper*, Yale University, New Haven, CT.

Raven, B. H. (1993). The basis of power: Origins and recent developments. *Journal of Social Issues, 49*(4), 227–241.

Rizkallah, E. G., & Miller, H. (2015). National versus private-label brands: Dynamics, conceptual framework, and empirical perspective. *Journal of Business & Economics Research, 13*(2), 123–136.

Robson, C. (1993). *Real world research*. Oxford: Blackwell.
Silverman, D. (2004). *Interpreting qualitative data: Methods for analysing talk, text and interaction* (2nd ed.). London: Sage Publications.
Stanković, L., & Končar, J. (2014). Effects of development and increasing power of retail chains on the position of consumers in marketing channels. *Ekonomika Preduzeća, 62*(5–6), 305–314.
Steiner, R. L. (2004). The nature and benefits of national brand/private label competition. *Review of Industrial Organization, 24*(2), 105–127.
Sutton-Brady, C., Kamvounias, P., & Taylor, T. (2015). A model of supplier–retailer power asymmetry in the Australian retail industry. *Industrial Marketing Management, 51*, 122–130.
Weitz, B., & Wang, Q. (2004). Vertical relationships in distribution channels: A marketing perspective. *Antitrust Bulletin, 49*(4), 859–876.
Yin, R. K. (2003). *Case study research: Design and methods* (3rd ed.). Applied Social Research Methods Series, Vol. 5. Thousand Oaks, CA: Sage Publications.
Yin, R. K. (2013). *Case study research: Design and methods* (5th ed.). Thousand Oaks, CA: Sage Publications.

2

The Management of FMCG Product Categories

Overview

This chapter examines the management of consumer goods product categories with a specific focus on food products and on the coexistence of private label and manufacturer brands. It explores category management and related topics in the marketing of consumer goods. Specific topics addressed are the history of category management, what category management is, the distinction between brand and category management, the history of private label, the deployment of the category management process, the objectives and benefits of category management, brand and category management organisational arrangements, research on category management and key aspects of the literature on the management of fast-moving consumer goods (FMCG) product categories.

R. Chimhundu, *Marketing Food Brands*,
https://doi.org/10.1007/978-3-319-75832-9_2

History of Category Management

Historical Development

Use of the term "category management" can be traced back to the year 1987 (Smith 1993), when a high profile FMCG company, Procter & Gamble, further developed its brand/product management system into a category management structure (Kracklauer et al. 2004; Mathews 1995). The early stages of use of the term were therefore from a manufacturer perspective of managing brands/products by category. This was before category management evolved to become more inclusive.

It is well documented that the development of category management originated in the USA with companies like Procter & Gamble and Coca-Cola (Hutchins 1997). It is not certain, however, whether there were no other companies practising category management. Some researchers (e.g. Hutchins 1997) have pointed out that a number of British companies such as Marks and Spencer may also have been managing their products by category at this time, although they may not have been using the term category management for their management technique. Furthermore, it is believed that some of the ideas and processes associated with category management may have been around for quite some time in one form or another in the packaged goods industry. What marked a difference from the old days was the significant investment in category management that was made by retailers and manufacturers in the 1990s (Dhar et al. 2001). The investment was coupled with a complete development and documentation of category management techniques in an unprecedented fashion. In this respect, category management is generally regarded to be a technique born of the 1990s, although one may reasonably assume that its roots date back to well before then. Additionally, on the retailer side, category management has been described as a "rediscovery" (Nielsen 1992).

The rediscovery argument is based on the fact that in the mid-1900s, US merchants (e.g. owners of local grocery stores) and customers often knew each other well as they lived in the same locality and communicated regularly on business and social matters. Shopkeepers often took and filled their customers' total orders. Retailers were therefore knowledgeable about their customers' requirements and tailored their merchandise and promotions accordingly. In return, the retailers were rewarded by their customers through shopping loyalty. Nielsen argues that today retailers are doing more or less the same thing through category management: "Retailers practising category management use information and technology to listen to their customers in much the same way their predecessors took advantage of their neighbourhood grapevine and over-the-counter chats with customers" (Nielsen 1992, p. 26).

Customised marketing and merchandising programmes are developed for individual categories based on information and technology. Thus, in a way the category manager (on the retailer side) is considered to be the new shopkeeper of retailing.

It is noted (Steiner 2001) that in the mid-1980s, pioneering vertical arrangements were made between Procter & Gamble and Walmart. The arrangements involved vertical or channel partnerships between a single retailer and a single manufacturer, and had the objective of realising benefits from the complementary functions performed by both retailers and manufacturers. Vertical or channel partnerships involved partnerships among organisations in the different stages of the supply chain. Soon, such vertical partnerships proliferated in the FMCG industry. In the 1990s, these individual channel partnerships were replaced by category management in supermarkets, discount stores and other outlets, as the number of individual partnerships had become a burden (Steiner 2001). It is recognised that Procter & Gamble's replacement of the brand/product manager system had a hand in facilitating the adoption of category management. However, the major impetus for the adoption of the practice is considered to have come from the supermarket industry. The industry had been losing market share to supercentres, discount stores, warehouse clubs and category killers (Steiner 2001). A category killer is a retail organisation that is dominant in a particular product category, and is so dominant to the extent of offering very low prices that smaller stores are not able to match. These mass merchant chains had taken advantage of trade wars that had separated manufacturers and supermarkets. Reacting to the loss of market share to mass merchants, the supermarket industry initiated action through its trade organisation, the Food Marketing Institute, and through its academic consultants as well. The action culminated in the production of five volumes of category management implementation plans (Blattberg 1995) to solidly formalise the "how to" of category management, even though the practice had started before the 1995 volumes.

Although historically category management is rooted in the food industry, it is expanding into other sectors. Since the mid-1990s when it started in the food industry, category management has swept across food and non-food categories in the USA, Europe and other countries such as Australia and New Zealand. In 1994, it had become fully operational in approximately 20% of major retailers in the US, and an additional 62% were in the development stage (McLaughlin and Hawkes 1994). Historically, the USA and European countries have been at the forefront of the development of category management, and it is expected that Australian and New Zealand retailers and manufacturers are followers in this area. Therefore, it would be reasonable to assume that different FMCG/supermarket environments would manage their categories in different ways.

Distinct Phases

Not much work in the academic literature has focused on a detailed chronological account of the history of category management. The account given in the preceding section has been constructed from various sources. The academic literature has also done little to analyse the historical development of category management. One article, though (Benoun and Helies-Hassid 2004), building on the work of Van der Ster (1993), has made an attempt to examine the history of category management from the viewpoint of the evolution of buyer–seller relationships, and has come up with four distinct phases. These phases are outlined as follows:

First Phase:

- Phase of pure/typical consumer marketing
- Initiated by manufacturers/suppliers
- Mass marketing techniques used
- Power in the hands of manufacturers/suppliers
- Retailers act as passive distribution points.

Second Phase:

- Phase of distributor/trade marketing
- Retailers interested in the consumer as well, and start to apply marketing techniques
- Products divided into categories and managed as such, but not in the full sense of category management as we know it today. The term "category management" is not even used
- Terms like "merchandising" and "key accounts" crop up
- Could argue that category management had started being applied but in a different way, using different terms.

Third Phase:

- Key account management has developed
- Manufacturers realise that it is necessary to target not only the consumer, but also the retailer/distributor. This becomes even more important when retailers centralise their organisations as the majority of purchasing decisions are centrally made
- Manufacturers start to consider retailers as strategic partners with whom close connections should be maintained. Individualised marketing activities target retailers
- The need for more cooperation between manufacturers and retailers becomes more pressing.

Fourth Phase:

- Phase of partnership, especially jointly initiated partnership
- Sharing of information that was previously confidential (between manufacturers and retailers)
- Category management as we know it today.

Although no dates have been given for the different phases, the general trend indicated in the phases partly reflects the earlier account given of the history of category management. A further analysis based on developments in France concluded that the French experience reflects three phases: the 1980s was the phase of trade marketing; the 1990s saw the emergence and development of ECR; and from the end of the 1990s to the beginning of 2000s there was a focus on category management to develop the demand side of ECR, which had been neglected in favour of the supply side (Benoun and Helies-Hassid 2004). This analysis partly reflects the general trend, except that not much information is given on what was happening before trade marketing. Overall, one would reason that the history of category management largely reflects a similar trend in most markets, but distinct characteristics are likely to exist from country to country. This may also have implications for the stage of development of category management in different markets, as well as the prevailing category management arrangements.

What is Category Management?

The term "category management" has been interpreted in various ways. At one point, a survey of British companies by the Institute of Grocery Distribution (IGD) in the mid-1990s came up with a number of definitions (Hutchins 1997). Possible reasons may have been that some companies did not quite understand what category management was, and that there were also varying perceptions of how it was supposed to be operationalised (Whitworth 1996). The early definition of category management had a manufacturer perspective; it came up in the late 1980s when Procter & Gamble, an FMCG giant, further developed its brand management structure into a category management structure. From this perspective, category management meant responsibility for managing multiple products (or product lines) rather than individual brands; and such responsibility fell in the hands of a category manager integral to the manufacturer organisation. According to Kotler (2000), this responsibility for managing manufacturer categories represented an additional layer of management above brand managers.

As time went on, the involvement of both manufacturer and retailer in the category management process became more pronounced in the definitions of category management, as shown by a 1995 definition of it as "a distributor/supplier process of managing product categories as strategic business units, producing enhanced business results by focusing on delivering consumer value" (Joint Industry Project 1995). Even then, research conducted by Hogarth-Scott and Dapiran (1997) in the Australian and UK food industries established that industry practitioners still had varying views of what actually constituted the concept of category management. The respondents, though, distinguished three broad areas: the categorisation process; availability of information as well as the sharing of such information in the channel; and partnership formation.

It is important for any research that is set in the category management context to clearly outline the operational definition of category management used in order to avoid confusion and differences in interpretation. In working towards this definition, a survey of the literature has identified a range of commonly used definitions. Table 2.1 outlines the respective elements of each.

In summary, category management is largely a joint process between manufacturers/suppliers and retailers. It can manifest itself in varying degrees of relationship/partnership/cooperation between the two. Product categories are central to category management; products are divided into categories that are then strategically managed. The strategic management of categories recognises each category as a business unit. Category management has a consumer focus, and focusing on consumer needs entails acquiring relevant information about the consumer and customising merchandising activities to the specific requirements of each store. Category management seeks to enhance business results for the retailer and the supplier/manufacturer through systematic buying/selling and merchandising, ensuring that those products that are demanded by the consumer are a priority and are made available on supermarket shelves. The three key stakeholders in the category management exercise therefore are retailers, manufacturers and consumers.

The operational definition of category management that is used in this book is modelled along the Nielsen (1992), Joint Industry Project (1995), Desrochers et al. (2003) and Kracklauer et al. (2004) definitions. Category management is a joint, manufacturer–retailer process of defining and managing product categories as strategic business units, focusing on satisfying consumer needs, and with the objective of producing enhanced business results.

Table 2.1 Definition of category management

Element(s)	Author(s)
Category Management	
Is a joint process (involves manufacturer/supplier and retailer)	Alvarazo and Kotzab (2001), Desrochers et al. (2003), Dupre and Gruen (2004), Joint Industry Project (1995), Joseph (1996)
Can be a manufacturer only or retailer only process	Kotler (2000), Lamb et al. (1996)
Involves partnership/cooperation/ sharing of information	Dussart (1996), Hutchins (1997), IGD (1999)
Involves mutual trust between manufacturer and retailer	Hutchins (1997)
Involves categorisation/product categories	Basuroy et al. (2001), Desrochers et al. (2003), Dupre and Gruen (2004), Dussart (1996), Hutchins (1997), Joint Industry Project (1995), Joseph (1996), Kracklauer et al. (2004), Nielsen (1992)
Is strategic management by product category (as strategic business unit)	Alvarazo and Kotzab (2001), Desrochers et al. (2003), Dupre and Gruen (2004), IGD (1999), Joint Industry Project (1995), Kracklauer et al. (2004), Nielsen (1992)
Entails coordination of buying and merchandising	Basuroy et al. (2001)
Has a consumer focus (consumer needs/consumer value)	Desrochers et al. (2003), Dupre and Gruen (2004), Dussart (1996), IGD (1999), Joint Industry Project (1995), Kracklauer et al. (2004), Nielsen (1992)
Involves customisation on a store-by-store basis	Alvarazo and Kotzab (2001), Desrochers et al. (2003), Joseph (1996), Nielsen (1992)
Seeks to enhance business results	Basuroy et al. (2001), Desrochers et al. (2003), Dupre and Gruen (2004), Dussart (1996), IGD (1999), Joint Industry Project (1995), Kracklauer et al. (2004)

Source: Created for this book based on the literature

The main elements worth noting therefore are:

- Product categories (categorisation)
- Strategic management by category
- Joint, manufacturer–retailer process
- Consumer focus, and
- Enhancement of business results.

Aspects of this definition will form the basis of the concepts that are studied in this book as far as the coexistence of manufacturer brands and private label is concerned.

Distinction between Brand Management and Category Management

Whereas category management represents strategic management by category, brand management represents strategic management by brand (Nielsen 1992). A category is a group of products that have a common consumer end use (Hofler 1996). The group of products is distinct, manageable and perceived by the consumer to be related and substitutable in meeting their needs (Blattberg 1995; Hogarth-Scott and Dapiran 1997). Examples of categories include dairy and frozen foods, soft drinks, household cleaners and paper products (Harris et al. 1999). The practice of defining and delineating the categories is not considered to be an easy task, as category boundaries are not always distinct and obvious (Johnson 1999); thus, there is a need for agreement between operating partners. Each category would normally also have smaller product groupings. The paper products category, for instance, could divide napkins, paper towels and toilet paper into distinct sub-categories (Nielsen 1992).

The terms "brand management" and "product management" are often used interchangeably. Traditionally, brand management entails responsibility for managing a single brand, although in reality it has now become increasingly common for a brand manager to be responsible for more than one brand (Homburg et al. 2000). Brands are created through branding and the essential function of branding is to create differences between need-satisfying offerings (Doyle 1993) in order to facilitate product identification (Lamb et al. 1996). A brand is therefore a "name, term, symbol, or combination thereof that identifies a seller's products and differentiates them from competitors' products" (Lamb et al. 1996: 297). A brand is a product that has dimensions that make it different from other products designed to satisfy a similar need (Keller 2003). Managing the brand is normally the responsibility of the brand manager, although it depends on the organisational set-up. In the academic and trade literature, there are no significant differences in the interpretation of what brand management stands for.

This study makes use of the following operational definition of brand management, adapted from Kotler (2000) and Hehman (1984): *Brand management is the coordination of marketing activities for a specific brand (product) that includes the development and implementation of the brand marketing plan and the monitoring of performance of the brand.*

Category management has largely served to systematically merchandise the many competing brands, especially manufacturer brands. However, manufacturer brands can be looked at as one type of brand as there are effectively two types of brands participating in FMCG/supermarket categories, namely private label brands and manufacturer brands. While manufacturer brands have historically dominated FMCG/supermarket categories, the growth of the private label (ACNielsen 2003, 2005), a brand that is owned and managed by the owners of the retail shelves, has meant that increased research attention can justifiably be focused on private label in relation to manufacturer brands in the product categories. In this respect, it is important to document the history of private label.

History of Private Label Brands

This account is largely based on five historical eras/landmarks identified in an academic article on the topic by Herstein and Gamliel (2004), and also integrates the works of other authors (e.g. Baden-Fuller 1984; Hoch and Banerji 1993; Hughes 1997; Veloutsou et al. 2004) into the five-era framework. Sketchy details of the historical account date as far back as 1840.

The time period that is historically considered to be the most significant regarding the emergence and development of private label brands is the modern marketing era (i.e. the mid-twentieth century); however, "an in-depth analysis of this phenomenon reveals that the era in which this branding approach became a huge commercial success began around the beginning of the twentieth century" (Herstein and Gamliel 2004: 63). Some authors (Hoch and Banerji 1993) have written about the turn of the twentieth century witnessing the "beginnings" of private label in grocery chains. This analysis is most likely from a commercial success point of view rather than being a categorical indication of the very beginnings of private label, since a number of authors have presented convincing accounts (with considerable detail and specific dates) of private label having been introduced for the first time in the nineteenth century (e.g. Fernie and Pierrel 1996; Herstein and Gamliel 2004).

The five main eras/landmarks in the history of private label have been identified by Herstein and Gamliel (2004) as:

- first era: emergence of private label brands (1840–1860);
- second era: the decline of private label brands (1861–1928);
- third era: the rise of private label brands (1929–1945);
- fourth era: the stability of private label brands (1946–1975);
- fifth era: private label brands versus manufacturer brands (1976–2003, or rather, to the present).

First Era: Emergence of Private Label Brands (1840–1860)

The first recorded evidence of private label dates as far back as 1840, when Jacob Bunn of Springfield, Illinois, in the USA, sold in his grocery shop certain items under his name or his family members' names. This was followed by a second grocer, Bernard H. Kroger, selling tea, coffee and pastries under his own name (Herstein and Gamliel 2004). There is agreement among the relevant authors (e.g. Hoch and Banerji 1993; Herstein and Gamliel 2004) that A&P (then the Great Atlantic & Pacific Tea Company) was the first grocery chain to sell private label brands. There is, though, a factual discrepancy in that Herstein and Gamliel (2004) claim that this happened in the 1860s while Hoch and Banerji (1993) state that this happened at the turn of the twentieth century. The first private label brands were therefore developed by entrepreneurial retailers. Manufacturer brands were sold alongside retailer and wholesaler brands. Consumers generally developed loyalty to their retailers who, from time to time, would grant them credit and accept barter.

Consequently, retailers wielded power in recommending what consumers would buy (Strasser 1989), and "it appears that because retailers were acting as advisers in the market place, they tended to persuade customers to buy unbranded goods and own-labels whose profit margins were mostly higher than manufacturer brands" (Herstein and Gamliel 2004: 64). In the UK, the first traces of private label brands also appeared on store shelves in the nineteenth century (Fernie and Pierrel 1996; Jeffery 1954; Key Note Market Review 2001), although the exact dates are not given. There is generally limited information on how private label started in Europe (Omana 2002).

From this account, one could say that the first private label brands originated in the USA and the UK (probably at an earlier date in the USA), and that manufacturer brands and private label have had a long history. As far as the period under consideration is concerned, the general lack of quantitative data makes it difficult to come to a conclusion as to how big private label was, in terms of market share, in relation to manufacturer brands.

Second Era: The Decline of Private Label Brands (1861–1928)

Against the background of private label that had gained ground in the USA, FMCG brands manufactured by companies such as Heinz, Procter & Gamble, National Biscuits, Coca-Cola, Quaker Oats, Colgate-Palmolive and Gillette suddenly experienced increased demand, especially after the American Civil War of 1861–1865 (Hoch and Banerji 1993). This occurred for a number of reasons (Herstein and Gamliel 2004). Firstly, in the USA the restoration of the country after the Civil War resulted in improved economic performance. The development of industries brought more disposable income, and together with the accelerated influx to urban life, there was more willingness to buy good-quality, innovative products in sophisticated packaging. Secondly, transportation systems improved, enabling manufacturer brands to be distributed nationally rather than just locally. Manufacturers learnt to become less dependent on wholesalers who had been responsible for distributing manufacturer products. Thirdly, national advertising in newspapers and magazines had a big impact. According to Hoch and Banerji (1993), manufacturer brands grew with each innovation in advertising media, and in addition, with the development of supermarkets.

Third Era: The Rise of Private Label Brands (1929–1945)

In 1929, the USA was hit by the Great Depression, ending a long period of wealth and economic prosperity. The depression lasted for a decade. Vast unemployment and low disposable incomes drove consumers to switch to private label brands because of their relative affordability in comparison to manufacturer brands.

Fourth Era: The Stability of Private Label Brands (1946–1975)

Manufacturer brands recovered soon after the Second World War due to factors such as economic improvement, the growth of a suburban middle class and heavy TV advertising (Herstein and Gamliel 2004). The post-war period experienced the diffusion of commercial television which dramatically changed the economics of mass communication and strengthened the role of manufacturer brands in food retailing (Fitzell 1982). By the middle of the twentieth century, private label had become a prominent element of merchandising policies for a number of UK retailers such as Sainsbury's and Marks and Spencer (Ogbonna and Wilkinson 1998). On the European continent in 1964, resale price maintenance (RPM) was abolished. This meant that retailers were no longer obliged to follow what manufacturers stipulated concerning selling prices. Private label brands therefore started to play an important role in the food sector (Herstein and Gamliel 2004). The end of RPM in the UK "heralded the launch of own label development on a large scale by many UK retailers" (Fernie and Pierrel 1996: 49). Its withdrawal contributed to a shift in power relations from food manufacturers to food retailers (Burns et al. 1983; Doel 1996). Consequently, in the mid-1960s, private label was increasingly perceived as a direct threat to manufacturer brands (Ogbonna and Wilkinson 1998).

In some respects, private label experiences differ from country to country. While this fourth era has been termed "the stability of retailer brands" (Herstein and Gamliel 2004), largely based on the experiences of certain European countries as well as those of the USA, it is noted that the situation in other countries was somewhat different. For instance, the first generic brands in New Zealand were introduced in the 1940s by Foodstuffs, a large grocery retail organisation, as a reaction to its inability to source some manufacturer brands (Keen 2003). However, the history of one of the major private label brands, Pams, is documented as dating back to 1937 when it was first introduced to enable the Four Square group of stores, which is part of Foodstuffs and has smaller stores than the larger supermarkets, to compete with larger chain stores (www.foodstuffs.co.nz). On another level, the Woolworths private label brand, which was introduced in 1969, claims to have been the first true private label brand in the country (Keen 2003). All these dates nevertheless show that, in global terms, countries like New Zealand are followers to countries such as the UK in terms of the emergence and development of private label.

Fifth Era: Private Label Brands Versus Manufacturer Brands (1976–2003)

In the mid-1970s, private label brands were generally of poor to mediocre quality and were perceived as being inferior to manufacturer brands (de Chernatony 1989). In both the UK and USA, the private label brands of the 1970s were mostly generics (Coyle 1978; McGoldrick 1984). They competed with manufacturer brands on the basis of price rather than quality, and innovation on the part of retailers was largely non-existent (Herstein and Gamliel 2004). In the UK in the 1970s, in reaction to the tough economic recession, supermarkets were forced to adopt a new strategy in order to invigorate trade. New lines of private label were introduced and offered at very alluring prices, thus considerably pressuring manufacturers to offer substantial discounts (Ogbonna 1989).

In the 1980s, private label brands in the UK and USA started changing to become increasingly targeted at the middle to upper consumer segments. This entailed producing upmarket products that competed with manufacturer brands on quality (Herstein and Gamliel 2004; Sansolo 1994). Private label therefore became more active in terms of product and packaging innovation. Yet in comparison terms, UK private label brands were actually more innovative than their US counterparts; according to Hughes (1997), "The main contrast was the way in which UK food retailers actually pioneered the development of some product categories which had not been developed before by manufacturers in the branded marketplace" (Hughes 1997, p. 172).

In both countries, the idea that guided most distributors in upgrading quality was that consumers were willing to pay more for manufacturer brands of better quality; therefore, too much emphasis on attractive price on the part of private label would not adequately sell such brands (Herstein and Gamliel 2004).

The new policy saw the continuation of innovation on a massive scale in the 1990s. There was a complete change by distributors in the management of their private label. Private label brands of the highest quality were presented to the market. The huge gap between manufacturer brands and private label was reduced significantly, and in some instances, private label brands were actually of higher quality than manufacturer brands, as was the case for some products marketed by the UK food chain, Sainsbury's (Herstein and Gamliel 2004). In fact, "As a result of this marketing approach, private brands, in the 1990s, mainly in the food sector, succeeded in gaining significant market shares and became a real threat to manufacturer brands" (Herstein and Gamliel 2004: 66). In the UK, while there was an emphasis on upgrading quality, most food retailers also sold price-based private label in addition to upmarket products, as a competitive response to the entrance of deep discounters (Burt and Sparks 2003).

Private label development is claimed to be stronger in the UK and in most Western European countries in comparison to the USA due to a number of factors, such as higher retail concentration (Hoch and Banerji 1993) and the associated power of large retail chains as a result of sheer scale; reinvestment of profits in initiatives such as central warehousing, centralised distribution systems and information technology; smaller national markets tending to favour fewer national competitors (Hoch and Banerji 1993); strategic management decisions such as maintaining a strong three-tier private label structure across the quality spectrum; and more advanced product and packaging innovation.

Conclusions that can be drawn from the historical development of private label include the importance of innovation and upgrading private label quality, as well as marketing/branding/advertising, in helping to push the private label brand to higher levels; the role played by retail consolidation and concentration in influencing the growth of private label brands; and the impact of economic performance in influencing the growth or decline of private label. Aspects related to product innovation and category marketing support (category development), as well as retail concentration, are important concepts that form the basis of this study of manufacturer brands and private label in FMCG/supermarket product categories.

Having examined the development of both private label and manufacturer brands, a discussion of the remaining aspects of category management follows.

Deployment of the Category Management Process

A category management framework was published in 1995 as the "how to" of category management (Joint Industry Project 1995). It outlines eight steps for the proper implementation of category management, which are:

- category definition,
- category role,
- category assessment,
- category scorecard,
- category strategies,
- category tactics,
- category/plan implementation, and
- category review.

In deploying the category management process, this eight-step approach (Joint Industry Project 1995) acts as a guide for companies both within and outside the FMCG sector. Some retailers and manufacturers follow it as is, but many have streamlined and customised it to suit their needs. For example, SUPERVALU employs five steps, Miller Brewing makes use of four, Big Y observes the original eight and CROSSMARK (a sales and marketing agency) employs five or six steps (ACNielsen et al. 2006). Desrochers et al. (2003) have actually identified and synthesised six category management process frameworks from the academic and trade literature. The reality, however, is that all these variations share a common overall process involving four macro-steps: data gathering, assessment, decision making and implementation (ACNielsen et al. 2006).

The category management process framework itself is beyond the scope of this volume. Relevant aspects of this study are derived from the framework, and these include merchandising aspects related to shelf space, shelf facings, shelf position and share, as well as the strategic decisions relating to such issues. These concepts are therefore central themes of this book, as they enable direct comparisons to be made between manufacturer brands and private label. There is a specific focus on the coexistence of manufacturer brands and private label in supermarket product categories with respect to these concepts, and largely from the viewpoint of how the two are balanced. Since manufacturers have their own brands that are competing (and cooperating) with private label, and retailers are largely seen in the literature as holding the balance of power in relation to manufacturers in the category management relationship, it is envisaged that the theories of power have an important part to play in the coexistence of the two types of brand.

Objectives and Benefits of Category Management

Category management is considered to have two main strategic objectives. Firstly, to define business units as product categories rather than individual brands or product lines, which is a shift from the brand management approach. Product decisions are based on category-level goals.

Secondly, to customise the marketing effort very closely to the shopping patterns of a locality (Dussart 1998). According to Dupre and Gruen (2004), "Category management theory posits that retail's sales and profits will be maximized by an optimal mix of brands, SKUs (stock-keeping units), and pricing that is determined from the perspective of the consumer and is based on historical sales data" (p. 445). The objectives of category management can be summarised into qualitative and quantitative goals (Kracklauer et al. 2004).

With respect to qualitative objectives, the key goal is customer retention, and it is achieved through assortment competence (i.e. the ability to decide on the right mix and variety of products to sell) and consumer orientation. Assortment competence is gained via the adoption of an up-to-date and demand-responsive product mix, optimal depth and optimal breadth. Consumer orientation is gained through clear and logical presentation. With respect to quantitative objectives, the key goal is increased earnings, and it is achieved through increased revenues and cost reductions. Increased revenue is gained through pushing products that are profitable and also through an increase in productivity per unit of retail space. Cost reductions are realised via product mix overhaul and reduction in capital investment (Kracklauer et al. 2004).

All three parties, manufacturers (suppliers), retailers and consumers, are meant to benefit from the category management exercise. Academic, empirical research has reported some positive results of category management implementation. Dhar et al. (2001) found that the implementation of category management is perceived by manufacturers as having a positive impact on category performance, and Basuroy et al. (2001) found that the implementation of category management has a positive effect on retailer profitability and prices. In addition to the academic, empirical research, a survey by the IGD (2002) in the UK showed that the majority of practitioners of category management believe that category management improves sales, market share, profitability, inventory levels, consumer understanding and trading relationships. Furthermore, exploratory research conducted in the Australian and UK grocery industries with manufacturers and retailers (Hogarth-Scott and Dapiran 1997) established in detail the specific benefits that accrue to manufacturers/suppliers, retailers and consumers.

Despite the reported benefits, some researchers have warned of potential negatives for consumers. For instance, lower prices are listed (Hogarth-Scott and Dapiran 1997) as a potential benefit to the consumer. Some academic researchers, however (Basuroy et al. 2001), have found that category management, as a profit-generating strategy, has the potential to foster price increases that could compromise consumer value. Furthermore, regarding manufacturers and retailers, since category management is a joint effort, it is expected that both parties will benefit from the enhanced business results achieved, and that a healthy balance prevails in the category management relationship between the two parties. However, given the increasing trend in retailer concentration and consolidation and the power of retailers, some researchers have argued that category management is nothing less than the translation of power acquisition by retailers over manufacturers (Dussart 1998; Randall 1994). This would seem to underline the importance of investigating category management arrangements in the process of assessing the power relationships in the management of the categories. It is expected that the power relationships between grocery retailers and suppliers/manufacturers would naturally reflect the power relationships in the management of the coexistence of manufacturer brands and private label.

Brand and Category Management Organisational Arrangements

Brand and Category Organisation for the FMCG Manufacturer

Organisation for category management on the manufacturer side has not always meant an arrangement that facilitates a formal, joint process with retailers. It may just mean a switch from brand-orientated to category-orientated organisation for managing the product categories of the manufacturer. For instance, when Procter & Gamble developed further the category structure from the brand management structure in 1989 (Katsanis and Pitta 1995), this was a manufacturer-only organisational arrangement that added a senior management layer (category manager) above brand managers in order to facilitate a focus on product categories in the process of managing brands (Kotler 2000). Before this type of reorganisation, brand managers would be fighting both external competition and internal brands. The category manager focuses on the product group and ensures that brand managers are not sabotaging each other (Lamb et al. 1996). Depending on the organisation, the responsibilities of the category manager may range from the coordination of brand management responsibilities, largely focusing on marketing activities only, to the coordination of brand management responsibilities, covering both marketing activities and trade/sales activities. The latter arrangement would mean a close working relationship with the sales organisation to the extent that selling would, in a way, become part and parcel of the category manager's responsibilities. Seen from this perspective of category management, a manufacturing company would have management responsibilities at brand/product and category levels as outlined in Table 2.2.

Table 2.2 Brand/product manager and category manager concepts for the manufacturer

Product level	Management level	Responsibilities
Brand/product	Brand manager (or product manager)	Develops marketing strategy for the brand. Positions the brand. Identifies target segments for the brand. Evaluates the effect of alternative marketing strategies on brand performance. Recommends changes to brand strategy. Has responsibility for brand performance.
Product category	Category manager	Evaluates existing products/product lines in the category. Evaluates the mix of new and existing products/product lines within the category. Considers the effects of additions and deletions. Has a greater responsibility to sell. Has overall responsibility for category performance.

Source: Adapted from Lamb et al. (1996)

It is important to note that the category management arrangement discussed above is not the one in which a manufacturer manages the whole category (inclusive of own and competitor brands) on behalf of or in partnership with a retailer, as that comes under joint category management.

Supermarket Retailer and Category Management Organisation

On the retailer side, category management also entails focusing on category performance rather than individual brand performance. Deployment of category management on the retailer side can be done without manufacturer input. In such circumstances, retailers would be making use of in-house information and technology, together with research and analyses purchased from third-party organisations (Nielsen 1992). The retailer would have its own category manager(s) (or "retail category manager(s)") in the organisational set-up. The category manager would largely be performing more of a total role as "a buyer, a merchandiser, a salesman, and a manager—all at the same time" (Nielsen 1992: 39). Although the deployment of retail category management can occur without manufacturer involvement, it has been argued that the benefits of category management are enhanced if there is collaboration between manufacturers/suppliers and retailers (Desrochers et al. 2003), the reason being that sharing resources such as information and technology with manufacturers is more cost effective and efficient than going it alone (Nielsen 1992).

Joint Category Management Arrangements

Manufacturer/Retailer Collaboration

Category management was introduced in the FMCG industry as a mechanism for managing relationships between retailers and manufacturers (Hogarth-Scott and Dapiran 1997). Joint category management entails a combined and coordinated effort between manufacturer and retailer to manage product categories as strategic business units, and is characterised by the sharing of information and technology resources between the two. Joint category management leverages the skills and resources of manufacturers and retailers, and cooperation/collaboration is key to such an arrangement. The two former adversarial parties, manufacturers and retailers, ideally would have to stop fighting over who obtains the most value added and start working together to maximise profit in any product group (Freedman et al. 1997). The two perform valuable complementary roles. Retailers bring to the table point-of-sale data, merchandising knowledge and total store consumer measures (Blattberg and Fox 1995), as well as making shelf space available for the display of products. Manufacturers bring to the table knowledge of consumer demographics, consumer motivations for buying, knowledge of market trends (Blattberg and Fox 1995), information on new product development trends (Nielsen 1992) and brand/product development capability; this includes both new product development and the updating of existing products/brands. Research indicates, however, that while there is such a business partnership between manufacturer and retailer, the two parties' objectives are never aligned. There will always be competition for a share of the surplus, and retailers tend to have stronger bargaining leverage in this regard (Freedman et al. 1997). Consequently, the theoretical construct of power can be said to be central in category management relationships, inclusive of the strategic decisions made on the nature of the coexistence of private label and manufacturer brands in the categories. Category management implies that retailers and manufacturers have to interact to create and manage strategies and operations for product categories, and not just individual brands (Dupre and Gruen 2004). To facilitate such interaction, a collaborative structure that integrates category functions and decisions should be adopted. Christopher (2005) and Nielsen (1992) have outlined the nature of the collaborative structure from the point of view of the changing roles of manufacturers and retailers. In the old roles that emphasise brand management, the manufacturer communicates through a sales representative and the retailer communicates through a buyer. In the new roles that emphasise category management, there is integrated communication involving more functions on either side.

Category Captain Arrangements

A category captain arrangement is a form of category management arrangement (Chimhundu et al. 2015; Lindblom and Olkkonen 2005). Although retailers may have category managers who handle the management of retail categories, the retailers recognise that they cannot possibly develop all the marketing skills necessary to cover their full range of products. The retailers therefore engage in arrangements that seek to "unleash the expertise" of suppliers (O'Keeffe and Fearne 2002: 299). A common category management arrangement in this respect in the FMCG/grocery retail sector is category captainship. A category captain (CC), or category leader, is a manufacturer, usually the dominant supplier in the category, which is empowered by the retailer to undertake management of the category on behalf of the retailer or in partnership with the retailer. This involves managing own (i.e. CC) brands and competitor brands (Desrochers et al. 2003). The typical arrangement is that the CC supplies resources and information in exchange for active participation in category planning, development and growth (Blattberg and Fox 1995). CC responsibilities also include deciding shelf arrangements, allocating shelf facings and recommending prices for both its own brands and those of competitors (Bush and Gelb 2005; Desrochers et al. 2003; Dewsnap and Hart 2004). Although the CC role is deemed to bring efficiencies to the respective category due to the greater resources and access to information, this kind of category management set-up has often been criticised over the possibilities of competitive collusion and competitive exclusion (Balto 2002; Desrochers et al. 2003).

The arrangement, however, does not mean that the CC has exclusive power over the category. In a way, the retailer is still in charge: "Gratification of category leadership is the retailer's decision and he may replace any category leader at any time" (Dussart 1998: 54). In the category management set-up, supermarket retailers largely take a strong interest in category growth and other measures of category performance, and therefore have the desire to see CCs grow the entire categories (Urbanski 2001). The extent of control and the exclusivity of decision control of the CC largely depends on the specific arrangement, and arrangements vary. The CC may be entrusted with all category decisions by the retailer. The retailer may be open to second opinions from other manufacturers in the category (termed "co-captains", "validators" or "consultants"). The retailer may also consult third-party advisers with no vested interests in the respective category (Desrochers et al. 2003).

According to Steiner (2001), CC arrangements can range from strong to weak, and this largely depends on the depth and breadth of decision responsibility of the CC, and on the availability and ability of other manufacturers to influence the decision. With respect to the different types of CC arrangements, ranging on a continuum from powerful to less powerful captains, the less powerful ones are those whose decisions are checked by the retailer and by other parties within the category (Lindblom and Olkkonen 2005). This situation therefore highlights the relevance of power and dependency in category management arrangements, and effectively in the nature of the coexistence of brands that are owned by the owners of the retail shelves (i.e. private label) and those that are owned by FMCG manufacturers (i.e. manufacturer brands).

Research Set in the Category Management Context

This book investigates the coexistence of manufacturer brands and private label in FMCG/supermarket product categories. Since the category management practice has become widespread in the FMCG/supermarket industry around the world (Dussart 1998), and manufacturer brands and private label largely coexist within this context, it becomes imperative that aspects of the category management set-up that are deemed to be relevant to the research on manufacturer brands and private label in grocery categories be taken into account in this study. The research is, however, not positioned primarily as category management research, but rather as research that investigates specific aspects of private label and manufacturer brands in FMCG/supermarket categories, taking into account relevant aspects of the category management set-up. A brief review of related research on the topic of category management follows, with a view to incorporating relevant aspects into the research direction of this volume.

An agenda for academic research on category management related issues was set by Hutchins (1997) in the process of trying "to raise awareness of category management among the academic community" (p. 180) and, at a later date, Dewsnap and Hart (2004) observed that although there is no shortage of coverage of category management related issues in the press, in trade publications and at practitioner conferences, the academic literature in the area is still limited. Broad areas of research agenda issues suggested by Hutchins (1997) that to date have not received adequate attention include the management of categories and related relationships, including power relationships within the categories. One such issue is that "Problem situations are bound to arise. Take for example, a situation when research indicates a retailer should delist his own-label [private label] product in favour of a competing brand. The solution to this is unclear" (Hutchins 1997: 179). The author contends that not only is the solution unclear, but the reactions and strategic approaches to such challenges may also not be standard across FMCG industries in different economies and across grocery retail chains. These aspects therefore can be better understood through studies that involve discussions with the relevant industry practitioners. There are power issues involved and power relationships between FMCG manufacturers and grocery retailers may vary. It is well known that in the FMCG industry the balance of power now rests with the retail chains, and one would assume that if coercive power was the dominant source of power used, decisions taken on such category issues as the one just given would be different from decisions that would be taken if other sources of power than coercive power were dominant in the category relationships.

With respect to manufacturer and retailer brands in FMCG categories, the facts at hand are as follows: Firstly, the two types of brands sit side-by-side on grocery retail shelves in a state of competition. Secondly, the technique of category management is being employed in some form, but this technique has not been adopted in a standard fashion by practitioners, so it is hard to tell with certainty the power relationships at play within the categories. Thirdly, the balance of power in manufacturer–retailer relationships is largely in the hands of grocery retail chains and the retail chains have the capacity to employ coercive power in their dealings with manufacturers/suppliers to their benefit, if they wish. It would be reasonable in this respect to expect that the more concentrated the grocery retail environment, the greater the power grocery retailers would have over manufacturers. This is partly because if a manufacturer brand is delisted by a grocery retail chain in a highly concentrated grocery retail environment, the manufacturer will have lost a huge chunk of its business, and this may actually threaten the survival of that manufacturer. Given this situation, power relationships between manufacturer brands and private label need to be better understood, especially in environments that have high grocery retail consolidation and concentration, as such environments are associated with even greater power on the part of grocery retail chains in relation to the FMCG manufacturers they deal with.

The research streams in the area of category management in the past two decades or so have tackled a number of areas. One stream has looked at profitability, category performance and assortment issues (e.g. Basuroy et al. 2001; Broniarczyk et al. 1998; Dhar et al. 2001; Gajanan et al. 2007; Zenor 1994). Another stream has addressed retailer–supplier relationships, relational outcomes, interaction and partnership (e.g. Dupre and Gruen 2004; Glynn 2007; Hogarth-Scott 1999; Lindblom 2001), and another has dwelt on marketing/brand management and sales organisational issues as they relate to category management (e.g. Dewsnap and Jobber 1999; Dussart 1998; Gruen and Shah 2000). One stream has looked at the historical development of category management (e.g. Benoun and Helies-Hassid 2004), and yet another has focused on research opportunities and/or extending the practice to other industries (e.g. Dewsnap and Hart 2004; Hutchins 1997).

A stream of investigation considered by the author to be much closer to this research and therefore more relevant has tackled power issues as they relate to the category management practice and category participant relationships, including the role of CCs, but this stream has fallen short of addressing in full such issues as they specifically relate to the coexistence of manufacturer brands and private label in the categories (e.g. Dapiran and Hogarth-Scott 2003; Desrochers et al. 2003; Hogarth-Scott and Dapiran 1997; Kurtulus and Toktay 2005; Lindblom and Olkkonen 2005; Lindblom and Olkkonen 2006). What seems to come out of this stream of literature, explicitly or implicitly, is that the retailers have the overall power in the category management set-up. It would be enlightening from an academic perspective to establish the role played by this power in the determination of the nature of coexistence between manufacturer brands and private label in the product categories. Therefore, whether the employment of such power would lean more towards coercive or non-coercive sources of power, and why, are issues that still have to be established through research. The author reasons that how these issues are treated is specific to grocery retail environments, grocery retail chains and the categories in question, since circumstances may differ between these units.

In addition, the types of research undertaken so far include empirical (quantitative, qualitative and a combination), experimental, historical, commentary and reviews. Lindblom and Olkkonen (2006) have noted, however, that there is a shortage of conceptual, qualitative empirical studies, and such studies are recommended. Glynn (2007) has noted that "much of the empirical work on category management has involved scanner data analysis rather than reporting retailer attitudes" (p. 63). The author of this work would posit that, given the power of retailers, retailer strategic thinking has a huge influence in determining the nature of coexistence between manufacturer brands and private label in FMCG/supermarket product categories.

Furthermore, the contextual environment of these studies undertaken so far is largely restricted to the USA and Europe. Australia has a minimal presence, and New Zealand has hardly featured in the academic journals, save for a few works such as Glynn (2007). It is recognised that context may not matter that much in academic research, unless there are characteristics that distinguish one context from another. In this case, distinguishing characteristics exist in New Zealand. The New Zealand FMCG/supermarket sector has the highest level of consolidation and concentration of all countries in the Organisation for Economic Co-operation and Development (OECD), and this should have an effect on power/dependency issues. Besides, academic research (Dapiran and Hogarth-Scott 2003) has hinted at the relationship between high retail concentration and readiness to employ certain bases of power such as coercive power. The New Zealand grocery sector would therefore offer fertile ground for a largely qualitative empirical study on the coexistence of manufacturer brands and private label.

Key Aspects of the Literature on the Management of FMCG Categories

The five-point definition of category management developed includes the following constructs: product categories (categorisation), strategic management by category, joint manufacturer (supplier)/retailer) processes, consumer focus and enhancement of business results. These issues have relevance to the conceptual framework on the study of the coexistence of manufacturer brands and private label in FMCG/supermarket product categories.

On the management of the categories, category management arrangements (specifically CC arrangements) and shelf matters (i.e. shelf space, shelf position, rationalisation and stocking/deletion decisions) also form part of the conceptual framework for this book.

From the historical development of private label, the importance of innovation and upgrading of private label quality was established, as well as the role of marketing/branding/advertising in helping to push the private label to higher levels and the role played by retail concentration in influencing the growth of private label. Retail consolidation and concentration, as well as product innovation and category marketing support (category development), are therefore relevant aspects that form an integral part of this research.

The power of grocery retailers, due in part to retail consolidation and concentration, was established as characterising the state of affairs in the FMCG/supermarket industry. This is seen as having implications for power relationships relating to the coexistence of manufacturer brands and private label in FMCG/supermarket categories. Power is therefore a relevant and appropriate topic of investigation in this study. Power relationships between manufacturer brands and private label need to be better understood, especially in environments that have high grocery retail consolidation and concentration.

There is a need for more academic research in the area of the management of FMCG categories and related relationships, including power relationships within the categories. In addition, contextual environments outside of Europe and the USA have been neglected in such research. The New Zealand grocery sector, with its highly concentrated grocery retail environment, offers such a research opportunity.

Chapter Recap

This chapter has explored the history of category management, what category management is, the distinction between brand management and category management, the history of private label, deployment of the category management process, objectives and benefits of category management, brand and category management organisational arrangements, streams of research in the category management context, and key aspects of the literature on the management of FMCG product categories. The next chapter covers product innovation, category marketing support, consumer choice and power issues in the FMCG industry.

References

ACNielsen. (2003). *The power of private label: A review of growth trends around the world*. New York, NY: ACNielsen.

ACNielsen. (2005). *The power of private label: A review of growth trends around the world*. New York, NY: ACNielsen.

ACNielsen, Karolefski, J., & Heller, A. (2006). *Consumer-centric category management: How to increase profits by managing categories based on consumer needs*. Hoboken, NJ: Wiley.

Alvarazo, U. Y., & Kotzab, H. (2001). Supply chain management: The integration of logistics in marketing. *Industrial Marketing Management, 30*(2), 183–198.

Baden-Fuller, C. W. F. (1984). The changing market share of retail brands in the UK grocery trade 1960–1980. In C. W. F. Baden-Fuller (Ed.), *The economics of distribution* (pp. 513–526). Milan: Franco Angeli.

Balto, D. (2002). Recent legal and regulatory developments in slotting allowances and category management. *Journal of Public Policy and Marketing, 22*(2), 289–294.

Basuroy, S., Mantrala, M. K., & Walters, R. G. (2001). The impact of category management on retailer prices and performance: Theory and evidence. *Journal of Marketing, 65*(4), 16–32.

Benoun, M., & Helies-Hassid, M. (2004). Category management, mythes et realites. *Revue Francaise Du Marketing, 198*(3/5), 73–86.

Blattberg, R. C. (1995). *Category management: Guides 1–5*. Washington, DC: Food Marketing Institute.

Blattberg, R. C., & Fox, E. J. (1995). *Category management: Getting started, guide 1*. Washington, DC: Food Marketing Institute.

Broniarczyk, S. M., Hoyer, W. D., & McAlister, L. (1998). Consumers' perceptions of the assortment offered in a grocery category: The impact of item reduction. *Journal of Marketing Research, 35*(2), 166–176.

Burns, J. A., McInerney, J., & Swinbank, A. (1983). *The food industry: Economics and policies*. London: Heinemann.

Burt, S. L., & Sparks, L. (2003). Power and competition in the UK retail grocery market. *British Journal of Management, 14*(3), 237–254.

Bush, D., & Gelb, B. D. (2005). When marketing practices raise antitrust concerns. *MIT Sloan Management Review, 46*(4), 73.

Chimhundu, R., Kong, E., & Gururajan, R. (2015). Category captain arrangements in grocery retail marketing. *Asia Pacific Journal of Marketing and Logistics, 27*(3), 368–384.

Christopher, M. (2005). *Logistics and supply chain management: Creating value-adding networks*. London: Financial Times/Prentice Hall.

Coyle, J. S. (1978, February). Generics. *Progressive Grocer, 28*(26), 75–84.

Dapiran, G. P., & Hogarth-Scott, S. (2003). Are co-operation and trust being confused with power? An analysis of food retailing in Australia and New Zealand. *International Journal of Retail and Distribution Management, 31*(5), 256–267.

de Chernatony, L. (1989). Branding in the era of retailer dominance. *International Journal of Advertising, 8*(3), 245–260.

Desrochers, D. M., Gundlach, G. T., & Foer, A. A. (2003). Analysis of antitrust challenges to category captain arrangements. *Journal of Public Policy & Marketing, 22*(2), 201–215.

Dewsnap, B., & Hart, C. (2004). Category management: A new approach to fashion marketing. *European Journal of Marketing, 38*(7), 809–834.

Dewsnap, B., & Jobber, D. (1999). Category management: A vehicle for integration between sales and marketing. *Journal of Brand Management, 6*(6), 380–392.

Dhar, S. K., Hoch, S. J., & Kumar, N. (2001). Effective category management depends on the role of the category. *Journal of Retailing, 77*(2), 165–184.

Doel, C. (1996). Market development and organizational change: The case of the food industry. In N. Wrigley & M. Lowe (Eds.), *Retailing, consumption and capital: Towards the new retail geography* (pp. 48–67). London: Longman.

Doyle, P. (1993). Building successful brands: The strategic options. *The Journal of Consumer Marketing, 7*(2), 5–20.

Dupre, K., & Gruen, T. W. (2004). The use of category management practices to obtain sustainable competitive advantage in the fast-moving-consumer-goods industry. *Journal of Business and Industrial Marketing, 19*(7), 444–459.
Dussart, C. (1996). EDI et management de categorie. *Decisions Marketing, 8*, 93–97.
Dussart, C. (1998). Category management: Strengths, limits and developments. *European Management Journal, 16*(1), 50–62.
Fernie, J., & Pierrel, F. R. A. (1996). Own branding in UK and French grocery markets. *Journal of Product and Brand Management, 5*(3), 48–59.
Fitzell, P. B. (1982). *Private labels, store brands and generic products*. Westport, CT: Avi Publishing Company.
Freedman, P. M., Rayner, M., & Tochlermann, T. (1997). European category management: Look before you leap. *McKinsey Quarterly, 1*, 156–165.
Gajanan, S., Basuroy, S., & Beldona, S. (2007). Category management, product assortment, and consumer welfare. *Marketing Letters, 18*(3), 135–148.
Glynn, M. S. (2007). How retail category differences moderate retailer perceptions of manufacturer brands. *Australasian Marketing Journal, 15*(2), 55–67.
Gruen, T. W., & Shah, R. H. (2000). Determinants and outcomes of plan objectivity and implementation in category management relationships. *Journal of Retailing, 76*(4), 483–510.
Harris, J. K., Swatman, P. M. C., & Kurnia, S. (1999). Efficient consumer response (ECR): Australian grocery industry. *Supply Chain Management, 4*(1), 35–42.
Hehman, R. D. (1984). *Product management*. Homewood, IL: Dow Jones-Irwin.
Herstein, R., & Gamliel, E. (2004). An investigation of private branding as a global phenomenon. *Journal of Euromarketing, 13*(4), 59–77.
Hoch, S. J., & Banerji, S. (1993). When do private labels succeed? *Sloan Management Review, 34*(4), 57–67.
Hofler, R. (1996). *Glossary of grocery industry terms*. Stamford: Progressive Grocer Associates.
Hogarth-Scott, S. (1999). Retailer-supplier partnerships: Hostages to fortune or the way forward for the millennium? *British Food Journal, 101*(9), 668–682.
Hogarth-Scott, S., & Dapiran, G. P. (1997). Shifting category management relationships in the food distribution channels in the UK and Australia. *Management Decision, 13*(4), 310–318.
Homburg, C., Workman, J. P., Jr & Jensen, O. (2000). Fundamental changes in marketing organisation: The movement toward a customer-focused organisa-

tional structure. *Journal of the Academy of Marketing Science,* 28(4), 459-478.

Hughes, A. (1997). The changing organization of new product development for retailers' private labels: A UK–US comparison. *Agribusiness, 13*(2), 169–184.

Hutchins, R. (1997). Category management in the food industry: A research agenda. *British Food Journal, 99*(5), 177–180.

IGD (Institute of Grocery Distribution). (1999). *Category management in action*. Watford: Institute of Grocery Distribution.

IGD (Institute of Grocery Distribution). (2002). *Category management: Which way now? Debate, evolution and future destination*. Watford: Institute of Grocery Distribution.

Jeffery, J. B. (1954). *Retail trading in Britain 1850–1950*. Cambridge: Cambridge University Press.

Johnson, M. (1999). From understanding consumer behaviour to testing category strategies. *Journal of the Market Research Society, 41*(3), 259–288.

Joint Industry Project. (1995). *Category management report.* Joint Industry Project on Efficient Consumer Response, USA.

Joseph, L. (1996). *The category management guidebook—Discount store news*. New York, NY: Lebhar-Friedman.

Katsanis, L. P., & Pitta, D. A. (1995). Punctuated equilibrium and the evolution of the product manager. *Journal of Product and Brand Management, 4*(3), 49–60.

Keen, E. (2003). Private Label Development. BCom Honours dissertation, University of Otago, Dunedin, New Zealand.

Keller, K. L. (2003). *Strategic brand management: Building, measuring, and managing brand equity* (2nd ed.). Upper Saddle River, NJ: Prentice Hall.

Key Note Market Review. (2001, March). *Own brands*. Hampton: Keynote Publications.

Kotler, P. (2000). *Marketing management* (10th ed.). Upper Saddle River, NJ: Prentice Hall.

Kracklauer, A. H., Mills, D. Q., & Seifert, D. (2004). Collaborative customer relationship management (CCRM). In A. H. Kracklauer, D. Q. Mills, & D. Seifert (Eds.), *Collaborative customer relationship management* (pp. 25–45). Boston, MA: Springer.

Kurtulus, M., & Toktay, L. B. (2005). Category captaincy: Who wins, who loses? *ECR Journal, 5*(1), 59–65.

Lamb, C. W., Jr., Hair, J. F., Jr., & McDaniel, C. (1996). *Marketing* (3rd ed.). Cincinnati, OH: South-Western.

Lindblom, A. (2001, December). *Institutionalisation of category management in the manufacturer-retailer relationships*. Paper presented at the ANZMAC Conference, Massey University, Palmerston North, New Zealand.

Lindblom, A., & Olkkonen, R. (2005, December). *Category captain arrangements in the Finnish grocery supply chain*. Paper presented at the ANZMAC Conference, University of Western Australia, Perth, Australia.

Lindblom, A., & Olkkonen, R. (2006). Category management tactics: An analysis of manufacturers' control. *International Journal of Retail and Distribution Management, 34*(6), 482–496.

Mathews, R. (1995). The power of category management. *Progressive Grocer, 74*(8), 12–14.

McGoldrick, P. (1984). Grocery generics: An extension of the private label concept. *European Journal of Marketing, 18*(1), 5–24.

McLaughlin, E. W., & Hawkes, G. F. (1994). *Category management: Current status and future outlook*. Ithaca, NY: Cornell University Food Industry Management Programme, Department of Agricultural Resource and Managerial Economics, Cornell University.

Nielsen. (1992). *Category management: Positioning your organisation to win*. Chicago, IL: NTC Business Books.

Ogbonna, E. (1989). Strategic changes in UK grocery retailing. *Management Decision Journal, 27*(60), 45–50.

Ogbonna, E., & Wilkinson, B. (1998). Power relations in the UK grocery supply chain: Developments in the 1990s. *Journal of Retailing and Consumer Services, 5*(2), 77–86.

O'Keeffe, M., & Fearne, A. (2002). From commodity marketing to category management: Insights from the Waitrose category leadership program in fresh produce. *Supply Chain Management: An International Journal, 7*(5), 296–301.

Omana, R. M. G. (2002). The growing industry of private labels. In *Food and agribusiness monitor*. Centre for Food and Agribusiness, University of Asia and the Pacific, Manila.

Randall, G. (1994). *Trade marketing strategies* (2nd ed.). Oxford: Butterworth.

Sansolo, M. (1994). Battle of the brands. *Progressive Grocer, 73*, 65.

Smith, K. (1993). No brand is too small. *Progressive Grocer, 72*(12), SS4–SS5.

Steiner, R. L. (2001). Category management: A pervasive, new vertical/horizontal format. *Antitrust, 15*(2), 77–81.

Strasser, S. (1989). *Satisfaction guaranteed: The making of the American mass market*. New York, NY: Basic Books.

Urbanski, A. (2001). Captains courageous: Category management techniques for grocery industry. *Supermarket Business, 56*(11), S3.
van der Ster, W. (1993, Oktober). Partnershipping de volgende fase in de relatie detaillist-leverancier. *Tijdschrift Roor Marketing* (pp. 10–15).
Veloutsou, C., Gioulistanis, E., & Moutinho, L. (2004). Own labels choice criteria and perceived characteristics in Greece and Scotland: Factors influencing the willingness to buy. *Journal of Product and Brand Management, 13*(4), 228–241.
Whitworth, M. (1996), Category management: Special report. *The Grocer*, Supplement, November 9.
Zenor, M. J. (1994). The profit benefits of category management. *Journal of Marketing Research, 3*(2), 202–213.

3

Product Innovation, Category Marketing Support, Consumer Choice and Power

Overview

This chapter examines the literature on important marketing and management subject areas related to the coexistence of private label and manufacturer brands in the consumer goods industry. Incorporated into the review are the product innovation and category marketing support activities of manufacturer brands and private label. Specifically, the chapter addresses product innovation, category marketing support, consumer choice and power in the FMCG industry.

Product Innovation

Meaning of Innovation

The concept of innovation is relevant to this book because the FMCG/supermarket product categories it focuses on are made up of many brands that are constantly engaged in innovation. In the marketing literature, the term "innovation" has largely been associated with new product

R. Chimhundu, *Marketing Food Brands*,
https://doi.org/10.1007/978-3-319-75832-9_3

Table 3.1 Definition of innovation

Element(s)	Author(s)
It can be a new product (i.e. goods, services or ideas)	Damanpour (1991), Kotler (1991), Lamb et al. (1996), Lundvall (1992), Nohria and Gulati (1996)
It can be a new technique/method	Lundvall (1992), Nohria and Gulati (1996)
It can be a new form of organisation	Lundvall (1992), Nohria and Gulati (1996)
It can be a new market/market opportunity	Lundvall (1992), Nohria and Gulati (1996)
It can be a new policy	Nohria and Gulati (1996)
It can be a new solution	Doyle and Bridgewater (1998)
It can be a new source of customer satisfaction/customer value	Barker (2002), Doyle and Bridgewater (1998)

Source: Created for this book based on the literature

development. Table 3.1 outlines the key elements of some of the definitions used in the marketing and management literatures.

The concept of innovation is indeed a broad concept (Avermaete et al. 2003), as can be seen from the variety of elements in the table. What all these authors are in agreement about, though, is that innovation is to do with newness. Although innovation also largely involves new products, it should not be seen as restricted to new product development or even product-related issues. There are other types of innovation such as process innovation, market innovation and so on. Market innovation, however, has areas of overlap with product innovation. Innovation encompasses a number of elements. This book only focuses on those domains that have something to do with manufacturer brand and private label innovation within FMCG product categories. To this effect, it is envisaged that innovation elements relating to product, packaging and branding (as well as those related to marketing/market and sales issues) will take priority. This book therefore makes use of the terms "product innovation" or "brand innovation" to mean the creation of new or modified (i.e. updated) products/brands. The scope therefore ranges from completely new, breakthrough concepts to minor adjustments (incremental changes) as they relate to manufacturer brands and private label.

A relevant, six-category framework (Booz et al. 1982) can act as an appropriate guide in this matter, to encompass all innovation activities envisaged. The framework identifies the following categories of new products: new-to-the-world products, as in new products that create an entirely new market; new product lines, as in new products that allow a firm to enter an established market for the first time; additions to existing product lines, as in new products that serve to supplement a company's established product lines (e.g. pack sizes, flavours and other varieties); improvements and revisions of existing products, as in new products that provide improved performance or greater value and replace existing products; repositionings, as in current products that are targeted at new markets or market segments; and cost reductions, as in new products that provide similar performance at lower cost. Aspects covered by this framework are relevant to this study. The range of innovations covered by the framework largely incorporate the relevant product, packaging, branding, marketing and sales aspects that would be expected in the product categories. In addition, it has been stressed that innovations involve commercialisation (Brassington and Pettitt 1997); so, in this book, any resources, skills and activities connected with commercialisation will be seen as part and parcel of the innovation.

Continuum of Innovation

Studies on innovation have used a number of alternative classifications of the concept. Two such classifications constitute what is termed the "continuum of innovation" (Brassington and Pettitt 1997), and are worth discussing. These are the typology of continuous, dynamically continuous and discontinuous innovation, and the typology of incremental and radical innovation.

Continuous, Dynamically Continuous and Discontinuous Innovation

A clear distinction has been made between the characteristics of each of these three types of innovation (Zairi 1995). Continuous innovation does not involve changing the ground rules of competition; it does mean improving standards under existing market conditions. Defining characteristics include (Zairi 1995): no requirement for changes in consumer behaviour (normally, readily accepted by the consumer); product can offer new features, but without disrupting established patterns of behaviour; and product provides the same function in the same way as usual. This is the most frequent form of innovation and carries a low level of risk. Examples of such innovations include FMCG product updates (Brassington and Pettitt 1997) such as new varieties (e.g. flavours and fragrances) (Zairi 1995).

Defining characteristics of dynamically continuous innovation include (Zairi 1995): use of new technology to serve a function that is established in the marketplace; consumers need to change and adapt their behaviour to the new product, which still performs a familiar function; price increases due to the new technology; resistance may occur if behaviour change required is substantial and if costs are high; and product is a significant innovation (Brassington and Pettitt 1997). Examples of such innovations include the CD-ROM for home PCs (Brassington and Pettitt 1997) and the evolution from manual typewriter to electric typewriter, and then to word processor (Zairi 1995).

Discontinuous innovation involves changing the ground rules of competition. Distinguishing characteristics include (Zairi 1995): product is completely new and disruptive, therefore requires consumers to establish new patterns of behaviour; it is a rare occurrence, probably once every decade; product may combine old and new technology to enable new tasks that were not performed previously; if accepted, product results in a radical reshaping of lives. Examples include the first PC (Zairi 1995) and the first video recorder (Brassington and Pettitt 1997).

Incremental and Radical Innovation

Studies on innovativeness have highlighted the distinction between radical and incremental innovation (Johannessen et al. 2001). Incremental innovation is associated with innovations within a paradigm. The focus is on incremental improvements. Radical innovation is associated with revolutionary innovations (Dosi 1982; Dewar and Dutton 1986). Discontinuities totally redefine the industry by creating new technological paradigms or regimes (Tushman and Romanelli 1985). Some authors, though (e.g. Damanpour 1991), have differed in their interpretation of the two concepts by recognising radical and incremental innovation as representing a continuum applicable even within the same paradigm. This book takes the view that, since there are two extremes on the radical–incremental continuum, the radical extreme end represents the highest degree of departure from current practice and the incremental extreme end represents the lowest degree of departure from current practice. Thus, radical innovation involving a new paradigm is taken to be more radical than Damanpour's (1991) radical innovation within the same paradigm. Subjective judgement can therefore be used to establish the positions of respective innovations on the continuum, depending on the magnitude of departure from current practice.

As argued by Hage (1980), innovations therefore vary on a continuum ranging from radical to incremental. In this respect, it is reasonable to consider the two typologies of continuous, dynamically continuous and discontinuous innovation and incremental and radical innovation to be largely similar in application because both are almost reflections of each other. The extreme end of continuous innovation represents the extreme end of incremental innovation and the extreme end of discontinuous innovation represents the extreme end of radical innovation. The Booz et al. (1982) six-category framework also bears some similarities to these two typologies in the sense that the six categories suggest a range on a continuum from incremental to radical innovation. For example, new-to-the-world products can be regarded as radical innovation; and additions of new pack sizes or flavours, as well as other improvements and cost reductions, represent incremental innovations.

Brand Innovation, Growth and Competitiveness

The importance of innovation for the growth, competitiveness and success of firms and economies is well documented in the academic and trade literature. In fact, as Guinet and Pilat put it, "Innovation is the heartbeat of OECD economies. Without it firms cannot introduce new products, services and processes. They find it hard, if not impossible to gain market share, reduce costs or increase profits. In effect, if the pulse of innovation is missing, firms quite simply die" (Guinet and Pilat 1999: 63).

New products tend to grow far more rapidly in sales than existing products, thereby providing a large boost for category and company growth (Brenner 1994). In addition, new product sales are an important innovation measure for many companies (Brenner 1994). Increased expenditure on innovation is also associated with the desire to grow.

The second annual global survey of top executives by the Boston Consulting Group (BCG) (2005) surveyed 940 executives, covering 68 countries and all major industries, the FMCG sector included. The survey found that 74% of the executives intended to increase spending on innovation in the year 2005: "The big reason for the almost unwavering support to innovation is of course growth. Fully 87 percent of the participants in our survey said that organic growth through innovation had become essential to success in their industry" (BCG 2005: 9).

Industries with an acute need for innovation were identified as hot spots, and such hot spots included consumer products and retail organisations. Consumer products companies (including FMCG blue-chips such as Procter & Gamble) and retail companies are currently facing an environment characterised by consolidation, rising commodity prices and increasing advertising expenditure. The two sectors had the highest proportion of respondents planning to increase innovation expenditure, at 79% (BCG 2005). Other factors that have also led to an emphasis on innovation in industry and academia include increased developments in globalisation, competition and technological advancement (McAdam 2005). The focus on innovation is considered to be universal; therefore, companies generally find it hard to outperform each other, no matter how much they invest. In view of the fact that competing companies are seen to be investing heavily, many companies are, as a result, unwilling to fall behind on their investment in innovation (BCG 2005).

Other writers have concurred that high rates of innovation are essential for growth and that businesses that do not innovate eventually become obsolete and die (Doyle and Bridgewater 1998; Hardaker 1998; Lindsay 2004; Robert 1995). In addition, innovation is seen as a source of competitive advantage (Denton 1999; Hastings and Healy 2001; Johne 1999; Rundh 2005). Both incremental and radical innovations are deemed important in achieving growth and competitiveness. However, although radical innovations are seen as difficult to achieve in the FMCG sector it is recognised that such innovations can actually open up new categories (BCG 2004).

Coming on to FMCG product categories, it was further revealed in a study by Booz et al. (1982) that more innovative categories achieve higher degrees of success than less innovative categories, which tends to further underline the link between innovation, growth and competitiveness. Furthermore, other writers have emphasised the need on the part of supermarket retailers to achieve growth and related business results for their entire categories through a combined effort with manufacturers (Urbanski 2001). While the mechanics of category management itself are geared towards achieving better business results that include growth in sales and profit for the categories, the innovative activities of brands can be seen as a driving force in the categories, which brings their importance into perspective.

Manufacturer Brand and Private Label Innovation in Supermarket Product Categories

Most supermarket product categories are made up of both manufacturer brands and private label. The importance of engaging in constant brand/product innovation in order to retain a strong market presence is emphasised in the literature (e.g. Johne 1999; Keller 2003; Kung and Schmid 2015). Manufacturer brands and private label are always engaged in some level of innovation ranging from incremental to radical. It is recognised that a lack of innovation and differentiation on their part can easily lead to category commoditisation and reduced consumer value (Schaafsma and Hofstetter 2005). One aspect of interest is the capacity of manufacturer brands and private label to innovate (Collins-Dodd and Zaichkowsky 1999) in view of the fact that the two are both in competition, and in a way, collaboration at the same time. Other aspects that are worth taking note of are the incentives for, and the nature and progress of (Desrochers et al. 2003), such innovation.

Capacity to Innovate

Capacity to innovate can be measured in terms of resources (e.g. facilities and technological, financial or human resources), skills/expertise (i.e. knowledge and proficiency), and the related innovation output (e.g. rate of innovation; quality and success of innovation). Use of the term "expert resources", however, to mean expertise or skills, is not uncommon in academic and management circles. Therefore, the single term "resources" for innovation can still be used individually to incorporate both skills/expertise and the other resources outlined above.

Experts hold different views on the state of the contributions to innovation by private label and manufacturer brands (Conn 2005). In the literature, grocery retailers have been seen to be boosting their capacity to innovate. According to this stream of literature (e.g. Lindsay 2004), major retailers are now creating new product development teams that rival those of manufacturing companies. European retailers such as Tesco and Aldi have long had that strength, while the USA's Walmart has, in the early 2000s, come up strongly. Other retailers have also made an effort to adopt new, distinctive private label strategies (Lee 1997). This view largely holds that private label brands have become masters of their own destinies on innovation and can do a lot on their own without having to deal with manufacturers.

Conversely, another stream of literature has considered retailers to be largely followers of manufacturers when it comes to brand/product innovation and related marketing activities that can be classified as category support (i.e. category development). This stream holds that manufacturer brands are leading the way and setting the standard on innovation. Historically, retailers have largely been followers of manufacturer brands on innovation (Aribarg et al. 2014; Coelho do Vale and Verga-Matos 2015; Hoch and Banerji 1993; Olbrich et al. 2016). Hoch and Banerji (1993) have shown that companies that are more experienced in product innovation tend to have a greater capacity to develop novel and complex products than those that are less experienced (Alegre et al. 2005). Manufacturers are generally considered to be more experienced than retailers in this area. In addition, it is noted that private label development is not backed by enough research and development money to generate breakthrough new products (Conn 2005). According to Conn (2005), "Some experts believe that the national brands and category leaders are setting the bar for innovation because they have the resources and ability to develop, test, sample, advertise, merchandise, and market innovative products" (p. 57).

With regard to these two points of view, this author takes a cautious view to suggest that the issue of whether it is the manufacturer brand or it is now the private label that is at the forefront in this respect, and why, depends on the research environment. Private label development is at different stages in different parts of the world, and its situation is economy and industry specific. The situation for the UK FMCG/supermarket industry is not expected to be the same as that of, say, Australia or New Zealand. Therefore, individual, in-depth assessments focused on the specific environments in question would yield much more reliable and accurate pictures than mere generalisations of manufacturer brands and private label that may not apply in a standard fashion across all FMCG industries. Besides, it has already been noted in the key aspects of the literature in the previous chapter that most research has been carried out focusing on the UK, European and US research environments. It would be insightful to establish what the other relatively neglected research environments with regard to manufacturer brands and private label in the developed world, such as New Zealand (from an academic rather than commercial perspective), have to offer in advancing the academic literature.

Incentives, Nature and Progress of Innovation

In the literature, it is recognised that the nature of innovation undertaken in FMCG product categories is more incremental (i.e. continuous) and less radical. Rudder (2003) examined food manufacturers' product innovations and identified a pattern of greater reliance on product adjustments as compared to new-to-the-world products. In addition, the innovative practice has been that manufacturers go in first with their brand innovations and private label brands follow. According to Duke (1998), collaboration between manufacturers and supermarket retailers in the area of new product development holds a lot of promise only in theory, with manufacturers having to contribute research, development and manufacturing expertise and retailers contributing intimate consumer knowledge from scanning data and loyalty schemes. This combination, ideally, should enable highly effective new product development. Ironically, manufacturers claim that retailers have taken "a parasitic approach to new product development, allowing manufacturers to take all the costs and risk of NPD, and then creating similar own-label equivalents of successful innovations" (p. 98). Other authors have concurred on the issue (Collins-Dodd and Zaichkowsky 1999; Harvey 2000; Ogbonna and Wilkinson 1998).

Furthermore, a survey of 100 UK brand managers found that 51% had their brands copied in one way or another by supermarkets, and 82% of them lost sales to imitators. This is seen as a potential disincentive for innovation. The imitative behaviour could be seen as exploitation of manufacturer innovative capacity by retailers, and could be testament to the fact that manufacturer brands have a combined superior capacity to innovate as compared to private label. The impact of such behaviour on the industry has been highlighted by Ogbonna and Wilkinson (1998):

> The main problem with own label brands identified by the manufacturers was the increasing pace at which retailers could design products identical to those of manufacturers. This, they claimed, was detrimental to the whole industry in the sense that it could reduce innovation. (p. 83)

Nevertheless, it would not be reasonable to assume that manufacturer brand innovative activities would stop because of private label imitation, since such innovations help manufacturer brands to successfully compete against each other as well. It is highly likely anyway that the more innovative manufacturer brands will still stand to benefit for as long as retailers derive benefits from them in turn, in which case the retailers would continue to need and accommodate them. One would further reason that manufacturer brands that are more innovative and supportive (marketing-wise) within the categories are viewed more positively by grocery retailers than those that are not, because the grocery retailers benefit more from them. Grocery retailers would therefore take this issue into account in working out how the private label and the manufacturer brand should coexist in the categories. It is an aspect of strategic dependency that will be investigated in this book through examining further literature from a different angle and charting an appropriate research direction.

With respect to innovation in grocery product categories, the consumer packaged-goods literature has largely portrayed manufacturer brand innovation in relation to private label as a competitive tool that is employed against private label, in addition to competing with other manufacturer brands. Verhoef et al. (2000) noted that superior brand innovation is a successful strategy in competing with private label, while Kumar and Steenkamp (2007b: 51) established from their study of "scores of categories in over 20 countries across the world, [that] private label success is 56 percent higher in categories with low innovation compared to categories with high innovation". Other research has similarly found that private label brands are more successful in categories experiencing low innovation by manufacturers (Coriolis Research 2002), and that FMCG manufacturers have managed to grab share in traditional strongholds of private label through innovation (Information Resources Inc. 2005). The literature therefore largely suggests that manufacturer brand innovation does have a negative impact on private label in FMCG categories. The alternative view, of manufacturer brand innovation having a positive impact on private label, has not been investigated in depth. The inherent interdependence between manufacturer brands and private label in the category management set-up justifies a case for investigating manufacturer brand innovation as an enhancer of private label in the categories. One may reason that if a positive impact exists, then it would most likely be factored by the powerful retail chains into policies and strategies that govern the nature of coexistence between their own brands and manufacturer brands in FMCG/supermarket categories. How important manufacturer brand innovation is in this regard, therefore, is a question that can be answered fully through further investigation; a research direction that, inter alia, forms part of this book.

It is beyond the scope of this book to directly deal with innovations involving retail operations. It is recognised, however, that retailers have done a lot in this area. For instance, Harvey (2000) looked at innovation and competition in the UK supermarket industry and identified as a big innovation on the retail side the replacement of wholesale markets with distribution centres. Similarly, Merrilees and Miller (2001a) studied the Australian supermarket industry and identified a series of radical innovations that have shaped the industry. More innovative supermarkets were found to enjoy greater competitive advantage. In addition, a study of retail pharmacies (Merrilees and Miller 2001b) identified key success factors for radical service innovation. Furthermore, there is also the whole host of technologies such as retailer scanner technology that the retailers themselves champion and finance.

Category Marketing Support

The term category marketing support (or simply category support) is used in this study to refer to any marketing activities, other than product innovation and its commercialisation, that help to develop and grow the FMCG/supermarket categories. Such marketing activities would include advertising (which is the major activity), sales promotion, merchandising support, monies/revenues paid to the retailers by suppliers, market research, branding/brand development and brand management. Since such activities largely occur around brands, category support is therefore largely achieved indirectly through brand support.

It is also noted that there are areas of overlap between product innovation and category marketing support. For instance, it has been observed that innovations involve commercialisation (Brassington and Pettitt 1997), and the commercialisation of innovations naturally involves marketing activities. For convenience, product innovation and category support are taken as two separate bins in the conceptual framework guiding this study, but where applicable, joint reference will be made to the two.

As FMCG product innovation and marketing go together, the research direction derived from the literature on innovation is extended to category marketing support as well. The line of enquiry on this topic will be how important category marketing support is in influencing policies that govern the coexistence between private label and manufacturer brands in the categories. Similarly, despite the fact that manufacturer brands support activities make them competitive against other manufacturer brands and against private label, the positive impacts of manufacturer brand category support on private label will be scrutinised. If there is a positive effect, one might argue that it would most likely be considered by the grocery retail chains in the determination of policies that govern the nature of coexistence between manufacturer brands and private label in the categories. How important manufacturer brand category support is is therefore a question that will be answered through investigation.

Also, since there is a synergistic relationship between innovation and strong branding (Merrilees 2003), frequent reference will be made to "product innovation and category support". The support for this approach is based on Merrilees' (2003) study that drew on various literatures in conducting an analysis of the strong brand and innovation paradox; a paradox that is based on the belief that pursuing a strong brand inhibits innovation. The analysis concluded that both strong branding and innovation should be jointly and simultaneously managed as there is a synergistic relationship between them. Low levels of innovation would tend to erode the power of the brand. Therefore, where performance on innovation is concerned, branding still has a part to play.

Consumer Choice

Issues related to consumer choice are considered here to be relevant to the coexistence of manufacturer brands and private label in FMCG/supermarket product categories, as the reviewed definition of category management conducted earlier has consumer focus as a key element. In addition, it is common knowledge that the consumer is the ultimate customer for both types of brands. Consumer choice issues can be looked at from the points of view of the consumer having the power to choose from competing brands, and/or the consumer needing choice in the form of selection.

The three key participants in supermarket product categories are retailers, manufacturers and consumers. Grocery retailers control the point of sale for both manufacturer brands and private label as the two types of brands are sold through retail stores. It has been suggested that there is a need to genuinely focus on the consumer because s/he "drives what happens in the category" (ACNielsen et al. 2006: 18). Consumers have the freedom to choose (Kaswengi and Diallo 2015; Nelson 2002; Olbrich et al. 2016); therefore, in the category set-up consumers do wield considerable power—the power of choice. In supermarket categories, consumers can choose from a range of manufacturer brands and products, and private label brands and products. Kumar and Steenkamp (2007a) have shown that consumers in any grocery category can be divided into groups such as brand buyers, private label buyers, random buyers and so on, with varying degrees of group switching depending on circumstances. Consumers' expectations and actions have a role in determining the success of private label alongside the expectations and actions of retailers and manufacturers (Hoch and Banerji 1993). Consumer choice therefore has a part to play in the dynamics between manufacturer brands and private label.

In addition, in today's environment, "more open and competitive markets, greater fragmentation and individualisation, and increased spending power have vastly increased the range of choices available to people" (Nelson 2002: 185). Fragmentation of consumer choice has meant that consumers want offers that are largely formulated for them individually (Hyman 2002). In order to offer such edited choice, there has been a move away from the old sea of merchandise approach in the store. The offering of edited choice helps retailers to develop brand loyalty and enables the effective execution of point-of-purchase display (Kessler 2004).

The desire to address consumer choice is further illustrated in the mechanics of category management. Sales and demographic data are continuously evaluated in order to determine purchasing patterns involving who buys what in a particular category, where, how often, and how they spend. Such data are used in combination with market research that reveals consumer insights (i.e. attitudes, characteristics and shopping behaviour), enabling the development of customised strategies for individual categories in specific supermarkets (ACNielsen et al. 2006; Nielsen 1992). This way, the requirements of consumers are addressed.

A further dimension to consumer choice involves how consumers cope with choice management among the competing offerings. Research conducted by Nelson (2002) found that the trusted brand plays an important role, but the price factor is also used by many consumers. Moderating factors can even come into play; for instance, interventions from pressure groups, academia and the media can influence consumption patterns (Nelson 2002). And as far as trust of the brand is concerned, it is important to note that this is more pronounced in non-commoditised categories than in commoditised categories.

In a situation where it would just be a matter of brands competing on the shelves, the issue of consumer choice would be pretty straightforward. The following issues can be said to have complicated the subject of consumer choice. Firstly, in the category management set-up, private label has been dubbed a sacred cow by some; the reason being that it is considered to be protected by retailers in the category management process (Major and McTaggart 2005), thus retailers have their own strategic objectives in this area and can be said to be employing specific strategic management regimes as they see fit. Secondly, CCs (usually dominant suppliers), who are appointed by retailers to take care of some categories, have been accused of engaging in activities that restrict competition, hence limiting variety and consumer choice (Desrochers et al. 2003). Thirdly, research by Ogbonna and Wilkinson (1998) found no evidence that manufacturers and retailers were strictly observing the maintenance of competition and consumer choice: "there was no evidence from our own research to suggest that they were pursuing anything other than their own perceived rational interests" (p. 84). However, the environment of this study was the UK, so it would not be appropriate to generalise to all other markets.

Despite these factors, it can be assumed that within supermarket product categories, there is a level to which manufacturers and retailers observe consumer choice issues. There may, though, be strategic objectives that render consumer choice issues secondary. But at the same time, retailers may not be able to afford to have overambitious objectives for their private label brands at the expense of manufacturer brands if that would compromise retailer performance. Ailawadi (2001) has stressed the need to have both manufacturer brands and private label in product categories for more positive business results. In the changing grocery retail landscape where retailers are becoming increasingly powerful due to retail consolidation and concentration, among other factors, an examination of the role still played by consumer choice in the coexistence of manufacturer brands and private label in FMCG product categories is necessary. It is also important to establish whether other strategic interests are interfering with consumer choice.

Power and the FMCG Industry

An Overview

Power in this context refers to the influence or control that a group, organisation or individual exerts upon the decisions, attitudes and behaviours of other groups, organisations or individuals (Hunt and Nevin 1974). Yet it is not necessary to limit the definition of power to groups, organisations and individuals only. Any phenomenon that has an influence on another phenomenon can be said to have a measure of power over it. Since the late 1950s, a number of writers have contributed to the literature on power from different perspectives, including (in order of date) French and Raven (1959), Emerson (1962), Hickson et al. (1971), Hunt and Nevin (1974), Patchen (1974), Pfeffer (1981), Naumann and Reck (1982), Sibley and Michie (1982), Mintzberg (1983), Gaski (1984), Diamantopoulos (1987), Toffler (1990), Biong (1993), Handy (1995), Dapiran and Hogarth-Scott (2003) and Hunt (2015). It should be noted that this list is not meant to be exhaustive in any way, and that it is only the key contributions judged to be more relevant to this book on the coexistence of manufacturer brands and private label in FMCG product categories that will be given closer attention. However, it is important to mention that, from the viewpoint of typologies of power, French and Raven's (1959) typology has achieved more prominence in the literature, and has been the most widely used, probably because it is more appealing.

Bases/Sources of Power

To many observers, power is the same as coercion, force or oppression, but such negative approaches to power are just one aspect of it (Duke 1998). Power can be conceptualised in terms of its sources, also known as bases of power. French and Raven (1959) identified five sources of power: coercive power, reward power, legitimate power, referent power and expert power. Initially, the theoretical development of these power bases was set in the context of interpersonal relations, but the bases have now achieved wide recognition in other contexts, such as interorganisational contexts, particularly in the analysis of marketing channels (Kasulis and Spekman 1980).

Reward power is based on the ability to mediate rewards or to remove any negative outcomes. The strength of reward power is related to the magnitude of the rewards that can be mediated or the magnitude of the negative outcomes that can be decreased. Coercive power is a type of power that is based on the ability to punish the other member for not complying. There is a similarity between coercive and reward power. Withholding rewards, for instance, could be seen as a form of punishment. Legitimate power is based on the belief that one party has the right to prescribe behaviour. Sources of legitimate power include cultural values, acceptance of social structure and designation by a legitimising agent (French and Raven 1959). Legitimate power can also be perceived from the point of view of traditional and legal legitimacy as power sources (Diamantopoulos 1987). Referent power has its basis in one party's desire to identify with the other (French and Raven 1959). The party that desires to associate with the other party therefore automatically puts the latter in a position of power. The desire can manifest itself in a demonstration of desirable behaviour to identify with, or from similarities in the characteristics of the two parties, or from the fact that the two have been closely associated over time (Heskett 1972). Expert power is based on the belief by one party that the other party has special knowledge or expertise in a given area (Cartwright 1965). It is recognised that one problem associated with expert power is that of its durability. The power of the expert can be reduced drastically if, once given, it provides the receiving party with the ability to operate without assistance (Stern and El-Ansary 1977).

The total power of a firm is a combination of several power bases. Positive and negative synergistic effects can result from the combination of these power bases (Stern and El-Ansary 1977); therefore, there are interlinkages among the various bases of power. Some researchers (e.g. Diamantopoulos 1987) have found it necessary to dichotomise the five power bases into coercive and non-coercive power, for the reason that it can be difficult to operationalise the various non-coercive power bases (i.e. expert, referent, legitimate and reward). It has also been found that the use of non-coercive power results in greater levels of satisfaction (on the part of the organisations receiving the influence) than the use of coercive power (Hunt and Nevin 1974; Wilkinson 1979).

The five bases of power are applicable to FMCG product category relationships relating to manufacturer brands and private label. By virtue of the radical shift in the balance of power from FMCG manufacturers to FMCG retailers (i.e. grocery retailers), especially in environments characterised by very high retail consolidation and concentration, the retailers have the capacity to employ coercive power to achieve certain outcomes if they wish. Therefore, it would be a reasonable assumption to make under these circumstances that grocery retailers would be more inclined to employ coercive power if there are benefits to be derived. Coercive power can come in the form of threats to delist brands or products from the category, or even actual deletion itself. Reward power could come in the form of allowing more shelf space or better shelf positions to certain manufacturer brands for compliance or in the form of appointing certain manufacturers to category captainship. Legitimate power would be at play if, for instance, there was an acceptance on the part of manufacturers that, since retailers own the supermarkets, they have the legitimate right to set the rules of operation and the rules of coexistence between manufacturer brands and private label. Referent power, on the other hand, would be at play if, say, a retailer wanted to create customer pull into the store using leading manufacturer brands, and also if benefits were to be derived from displaying their own private label brands next to such manufacturer brands. Expert power would be at play if, for instance, manufacturer brands were seen as superior to private label brands on innovation and brand/category support and if such factors were critical for category development and had an effect on consumer choice.

It is important to closely examine how sound French and Raven's (1959) classification is before employing it, to ensure that any possible gaps that might arise in the analysis of power issues relating to manufacturer brands and private label are closed beforehand. On this topic, one of the later classifications (Mintzberg 1983) is made use of for comparison purposes. Mintzberg (1983) identified resources, technical skill, knowledge, formal power and access to others as sources of power. Yet technical skill and knowledge mean the same thing as French and Raven's expert power, while formal power can easily be equated to their description of legitimate power. Access issues can be linked to French and Raven's legitimate and/or reward power, and resources can easily be linked to their expert power (i.e. from the viewpoint of expert resources) and reward power (from the viewpoint of having the resources to reward). However, as far as the construct "resources" is concerned, there is still a gap in French and Raven's typology in the sense that physical resources such as machines, equipment, facilities and related resources do not quite fit into this typology and such resources have relevance for this book. Therefore, although French and Raven's (1959) classification is to a large extent water-tight, it can still be complemented by making reference to Mintzberg's (1983) "resources" as a power base, where it is deemed that there is a need to be more inclusive in theoretically describing the bases of power at play in this book. In addition, suggestions by Raven (1993) that information power should be taken as a sixth source have been challenged by other researchers (e.g. Dapiran and Hogarth-Scott 2003), who perceive it as being an aspect of expert power; a view that this author shares. In connection with this, with respect to activities such as brand/product innovation and category support, and issues such as consumer choice that have been discussed already in relation to the grocery retail categories, aspects related to the bases of power are deemed to be relevant in assessing such activities and their impact in shaping the coexistence of manufacturer brands and private label in FMCG/supermarket product categories.

Power–Dependence Relationship

The concept of dependence is central to most discussions of power (Diamantopoulos 1987). Relationships between parties usually imply mutual dependence between them, and power resides implicitly in one party's dependence on the other. Dependency theory argues that the power to influence or control rests in the extent to which one party depends on the other for the things that they value (Emerson 1962; Hickson et al. 1971; Pfeffer 1981). That dependence can manifest itself in economic, social or other ways. A balanced power relationship is a situation in which both parties have equal amounts of power (Hamlin and Chimhundu 2007; Wilkinson 1973). An asymmetrical power relationship exists where there is disparity in power between two or more parties (Diamantopoulos 1987). The party over which the power is held can also have countervailing power, and this is the ability to counter or neutralise the power of the other party (Hunt and Nevin 1974). The concept of countervailing power has been used by Howe (1990) to characterise the relationship between retailers and manufacturers in the UK grocery industry, and this may well apply to manufacturer–retailer relationships in other parts of the world.

Important to note also, according to Ogbonna and Wilkinson (1998), is the fact that relationships such as manufacturer–retailer relationships do not exist in a vacuum, but in a wider political and social context that regulates those relationships. There may be a need to observe fair trading conventions or to respect consumer sovereignty. The power of an organisation therefore is not necessarily only a function of the economic structure. Similarly, this research suggests that industry contexts would also be relevant. For instance, relationships set in low retail consolidation and concentration environments would be expected to be different from relationships set in high retail consolidation and concentration environments.

The earlier discussion on category relationships between manufacturers and retailers looked at a situation where either party brings something that is valued by the other party to the table. Therefore, despite the fact that it is widely recognised in the literature that power in the FMCG sector has largely shifted from manufacturers to retailers because of retail consolidation, retail concentration, the growth of retailer brands and the increased utilisation of information and technology by retailers (e.g. Berthon et al. 1997; Hogarth-Scott 1999; Nielsen 1992; Panigyrakis and Veloutsou 2000), power–dependence relationships still have a role to play on both sides. In FMCG product categories, "manufacturers have critical mass and specialisation, superior brand quality, variety, image and choice in many markets. They gain power from advertising and product innovation investments in brand equity" (Hogarth-Scott 1999: 670).

Retailers therefore depend on manufacturers in some ways; for instance, they rely on manufacturers bringing about and promoting superior offerings that attract consumers to the categories. There are aspects of strategic dependency in the categories.

Dependency within the categories can also be conceptualised from the viewpoint of types of interdependence. Thompson (1967) identified pooled, sequential and reciprocal interdependence as three forms of interdependence. Although these were conceptualised as types of internal interdependence, they can be easily applied to manufacturer–retailer category relationships, and specifically to manufacturer brands and private label. Pooled interdependence is a situation in which each unit "renders a discrete contribution to the whole and each is supported by the whole" (Thompson 1967: 54). If each party does not perform adequately, the whole organisation's performance is negatively affected. Sequential interdependence takes a serial form; the output of one unit acts as the input of another. There is direct interdependence between the units, but the interdependence is not symmetrical. Reciprocal interdependence involves give and take; it is a two-way relationship where the output of one unit is the input of the other, and vice versa. It is important to note that an organisation that has reciprocal interdependence automatically has sequential and pooled interdependence. An organisation that has sequential interdependence also has pooled interdependence. However, having pooled interdependence does not necessarily mean that the other two types of interdependence are present (Thompson 1967). With respect to grocery retail categories, it would be reasonable to say that if there is an element of category participants bringing something to the categories that other participants in turn would draw from, then there is an element of pooled interdependence.

Power and Cooperation

Retailers hold the balance of power in FMCG manufacturer–retailer relationships. Power has often been associated with the negative use of it by the party holding the balance of power. Punishment and conflict have often been interpreted as being synonymous with power. In the literature, employment of the concepts of power, dependence and cooperation has ranged from the use of it in the negative sense (largely coercive) to its employment in a more positive way that promotes cooperation. Dapiran and Hogarth-Scott (2003) emphasised that cooperation is achieved through the use of referent and expert power. The use of non-coercive power results in the strengthening of exchange relationships and increases trust (Hunt and Nevin 1974). The use of coercive power weakens relationships, reduces trust and tends to invite retaliation (Raven and Kruglaski 1970). Power should therefore not be perceived as something that is strictly contrary to cooperation, as such an understanding of power is considered to be rather narrow. The concepts of power and cooperation should not necessarily be seen as opposite concepts (Dapiran and Hogarth-Scott 2003). This book therefore takes note of the misinterpreted relationship between power and cooperation.

Balance of Power in the FMCG Industry

Retailers and manufacturers are key participants in the FMCG industry. Traditionally, successful FMCG manufacturers have wielded power over retailers, in many instances dictating marketing terms for their brands. In the 1960s and 1970s, manufacturers had a considerable influence on retailers' decisions relating to the stocking, displaying and pricing of manufacturer brands (Weitz and Wang 2004), but that is now changing in countries like New Zealand, Australia, the UK and the USA.

The changing FMCG landscape has witnessed a rise in the power of the trade. Since the 1980s, there has been a shift in power from manufacturers to the large grocery retail chains that now dominate the trade (ACNielsen et al. 2006; Berthon et al. 1997; Hogarth-Scott 1999; Hollingsworth 2004; Hovhannisyan and Bozic 2016; Kumar and Steenkamp 2007a, b; Panigyrakis and Veloutsou 2000; Stanković and Končar 2014; Sutton-Brady et al. 2015; Weitz and Wang 2004). Grocery retail chains no longer simply take instructions from manufacturers; they have become more actively involved in marketing (Kessler 2004). The rise in the power of supermarket retailers has been attributed to the combined effect of several factors: retail consolidation and concentration and the resultant sheer size/scale of retailers, the growth of private label, the increased utilisation of information and communication technology by retailers, and retailer approach to marketing (Berthon et al. 1997; Hogarth-Scott 1999; Hollingsworth 2004; Panigyrakis and Veloutsou 2000; Weitz and Wang 2004). Also, power retailers have largely become gatekeepers of the supermarket shelves (ACNielsen et al. 2006). From a retail concentration point of view, in a country such as New Zealand for instance, it would be disastrous for a brand to be delisted by either of the two major grocery retail players which in combination command a 98% or so share of the supermarket industry. In such an environment, it is to be expected that coercive power would be the dominant source of power in grocery retail and FMCG manufacturer relationships if the retail chains benefit in the process.

A look at some empirical studies in the literature that were based on the work of French and Raven (1959) uncovers research that challenges the assumption that buyer–seller relationships are dominated by negative power sources (e.g. Biong 1993; Naumann and Reck 1982; Patchen 1974; Sibley and Michie 1982). In terms of geographical environment, however, most of these studies were carried out in the USA, an environment with a comparatively low retail concentration level, and the studies were not really directly focused on the relationships as they relate to manufacturer brands and private label. A much later study by Dapiran and Hogarth-Scott (2003) conducted interviews in the UK and Australian FMCG industries, and based on analysis of the interviews, suggested among other things that, where there is high retail concentration and low grocery retailer dependence on the supplier, retailers are more likely to employ coercive power. In addition, it is noted that

> suppliers are still very much aware of the reward/coercive power resources of the retailers. And it is also obvious that the retailers are conscious of these power bases that could be used in the event that the suppliers decide not to cooperate. (p. 265)

While power has been used in studies of manufacturer and retailer relationships, the theory has not been employed to specifically focus on the coexistence of manufacturer brands and private label, especially in a changing and radically altered FMCG landscape characterised by high retail consolidation and concentration, as well as direct competition between manufacturer brands and private label, and where private label is still in a state of development and the respective powerful retail chains are emphasising private label development. Such is the environment as it relates to New Zealand, which has a much higher retail consolidation and concentration level than the UK and Australia; and one wonders whether in this set-up, the dominant sources of power are coercive or non-coercive, and why? It is therefore envisaged that there is a power–dependence relationship involving manufacturer brands and private label whose nature still has to be better understood through empirical research.

Chapter Recap

This chapter has covered product innovation, category marketing support and consumer choice, as well as power and the FMCG industry. The next chapter explores private label and manufacturer brand coexistence in the product categories.

References

ACNielsen, Karolefski, J., & Heller, A. (2006). *Consumer-centric category management: How to increase profits by managing categories based on consumer needs*. Hoboken, NJ: Wiley.

Ailawadi, K. L. (2001). The retail power performance conundrum: What have we learned? *Journal of Retailing, 77*(3), 299–318.

Alegre, J. N., Chiva, R., & Lapiedra, R. (2005). A literature-based innovation output analysis: Implications for innovation capacity. *International Journal of Innovation Management, 9*(4), 385–399.

Aribarg, A., Arora, N., Henderson, T., & Kim, Y. (2014). Private label imitation of a national brand: Implications for consumer choice and law. *Journal of Marketing Research, 51*(6), 657–675.

Avermaete, T., Viaene, J., Morgan, E. J., & Crawford, N. (2003). Determinants of innovation in small food firms. *European Journal of Innovation Management, 6*(1), 8–17.

Barker, A. (2002). *The alchemy of innovation: Perspectives from the leading edge.* London: Spiro Press.

BCG (Boston Consulting Group). (2004). *Breakthrough insights: Expanding categories, exposing needs.* Boston, MA: The Boston Consulting Group.

BCG (Boston Consulting Group). (2005). *Innovation 2005.* Boston, MA: The Boston Consulting Group.

Berthon, P., Hulbert, J. M., & Pitt, L. F. (1997). Brands, brand managers and the management of brands: Where to next? *MSI Report No. 97-122.* Cambridge, MA: Marketing Science Institute.

Biong, H. (1993). Satisfaction and loyalty to suppliers within the grocery trade. *European Journal of Marketing, 27*(7), 21–38.

Booz, Allen & Hamilton. (1982). *New product management for the 1980s.* New York, NY: New York Press.

Brassington, F., & Pettitt, S. (1997). *Principles of marketing.* London: Pitman Publishing.

Brenner, M. S. (1994). Tracking new products: A practitioner's guide. *Research Technology Management, 37*(6), 36–40.

Cartwright, D. (1965). Influence, leadership, control. In J. G. March (Ed.), *Handbook of organisations* (pp. 1–47). Chicago, IL: Rand McNally.

Coelho do Vale, R., & Verga-Matos, P. (2015). The impact of copycat packaging strategies on the adoption of private labels. *Journal of Product & Brand Management, 24*(6), 646–659.

Collins-Dodd, C., & Zaichkowsky, J. L. (1999). National brand responses to brand imitation: Retailers versus other manufacturers. *Journal of Product and Brand Management, 8*(2), 96–105.

Conn, C. (2005). Innovation in private label branding. *Design Management Review, 16*(2), 55–62.

Coriolis Research. (2002). *Responding to private label in New Zealand.* Auckland: Coriolis Research.

Damanpour, F. (1991). Organisational innovation: A meta-analysis of effects of determinants and moderators. *Academy of Management Journal, 34*(3), 555–590.

Dapiran, G. P., & Hogarth-Scott, S. (2003). Are co-operation and trust being confused with power? An analysis of food retailing in Australia and New Zealand. *International Journal of Retail and Distribution Management, 31*(5), 256–267.

Denton, D. K. (1999). Gaining competitiveness through innovation. *European Journal of Innovation Management, 2*(2), 82–85.

Desrochers, D. M., Gundlach, G. T., & Foer, A. A. (2003). Analysis of antitrust challenges to category captain arrangements. *Journal of Public Policy & Marketing, 22*(2), 201–215.

Dewar, R. D., & Dutton, J. E. (1986). The adoption of radical and incremental innovations: An empirical analysis. *Management Science, 32*(11), 1422–1433.

Diamantopoulos, A. (1987). Vertical quasi-integration revisited: The role of power. *Managerial & Decision Economics, 8*(3), 185–194.

Dosi, G. (1982). Technological paradigms and technological trajectories: A suggested interpretation of the determinants and directions of technical change. *Research Policy, 11*(3), 147–162.

Doyle, P., & Bridgewater, S. (1998). *Innovation in marketing*. Oxford: Butterworth-Heinemann.

Duke, R. (1998). A model of buyer-supplier interaction in UK grocery retailing. *Journal of Retailing and Consumer Services, 5*(2), 93–103.

Emerson, R. (1962). Power—Dependence relations. *American Sociological Review, 27*(2), 31–41.

French, J. R. P., & Raven, B. (1959). The bases of social power. In D. Cartwright (Ed.), *Studies in social power*. Institute of Social Research (pp. 150–167). Ann Arbor, MI: The University of Michigan.

Gaski, J. F. (1984). The theory of power and conflict in channels of distribution. *Journal of Marketing, 48*(3), 9–29.

Guinet, J., & Pilat, D. (1999). Promoting innovation: Does it matter? *The OECD Observer*, Paris, No. 217/218, pp. 63–65.

Hage, J. (1980). *Theories of organizations*. New York, NY: Wiley.

Hamlin, R. P., & Chimhundu, R. (2007). Branding and relationship marketing within the trifecta of power: Managing simultaneous relationships in consumer goods marketing. *Journal of Customer Behaviour, 6*(2), 179–194.

Handy, C. (1995). *Gods of management: The changing work of organizations*. Oxford: Oxford University Press.

Hardaker, G. (1998). An integrated approach towards product innovation in international manufacturing organizations. *European Journal of Innovation Management, 1*(2), 67–73.

Harvey, M. (2000). Innovation and competition in UK supermarkets. *Supply Chain Management: An International Journal, 5*(1), 15–21.

Hastings, K., & Healy, M. (2001, December). *The innovation chain: A strategic approach to gaining competitive advantage*. Paper presented at the ANZMAC Conference, Massey University, Palmerston North, New Zealand.

Heskett, J. L. (1972). Interorganisational problem solving in a channel of distribution. In M. Tuite, R. Chisholm, & M. Radnor (Eds.), *Interorganisational decision making*. Chicago, IL: Aldine Publishing Company.

Hickson, D. J., Hinings, C. R., Less, C. A., Schneck, R. E., & Pennings, J. M. (1971). A strategic contingencies theory of organisational power. *Administrative Science Quarterly, 16*(2), 216–229.

Hoch, S. J., & Banerji, S. (1993). When do private labels succeed? *Sloan Management Review, 34*(4), 57–67.

Hogarth-Scott, S. (1999). Retailer-supplier partnerships: Hostages to fortune or the way forward for the millennium? *British Food Journal, 101*(9), 668–682.

Hollingsworth, A. (2004). Increasing retail concentration: Evidence from the UK food sector. *British Food Journal, 106*(8), 629–638.

Hovhannisyan, V., & Bozic, M. (2016). The effects of retail concentration on retail dairy product prices in the United States. *Journal of Dairy Science, 99*(6), 4928–4938.

Howe, W. S. (1990). UK retailer vertical power, market competition and consumer welfare. *International Journal of Retail and Distribution Management, 18*(2), 16–25.

Hunt, S. D. (2015). The bases of power approach to channel relationships: Has marketing's scholarship been misguided? *Journal of Marketing Management, 31*(7–8), 747–764.

Hunt, S. D., & Nevin, V. R. (1974). Power in a channel of distribution: Sources and consequences. *Journal of Marketing Research, 11*(2), 186–193.

Hyman, R. (2002, Summer). The retail roller-coaster: Message from the High Street. *Market Leader*.

Information Resources, Inc. (2005, November). Private label: The battle for value-oriented shoppers intensifies. *Times and trends: A snapshot of trends shaping the CPG industry* (pp. 1–23).

Johannessen, J., Olsen, B., & Lumpkin, G. T. (2001). Innovation as newness: What is new, how new, and new to whom? *European Journal of Innovation Management, 4*(1), 20–31.

Johne, A. (1999). Successful market innovation. *European Journal of Innovation Management, 2*(1), 6–11.

Kasulis, J. J., & Spekman, R. E. (1980). A framework for the use of power. *European Journal of Marketing, 14*(4), 21–32.

Kaswengi, J., & Diallo, M. F. (2015). Consumer choice of store brands across store formats: A panel data analysis under crisis periods. *Journal of Retailing and Consumer Services, 23*, 70–76.

Keller, K. L. (2003). *Strategic brand management: Building, measuring, and managing brand equity* (2nd ed.). Upper Saddle River, NJ: Prentice Hall.
Kessler, C. (2004). Editorial: Branding in store: Marketing in the 21st century. *Journal of Brand Management, 11*(4), 261–264.
Kotler, P. (1991). *Marketing management: Analysis, planning, implementation and control*. London: Prentice Hall.
Kumar, N., & Steenkamp, J. E. M. (2007a). *Private label strategy: How to meet the store brand challenge*. Boston, MA: Harvard Business School Press.
Kumar, N., & Steenkamp, J. E. M. (2007b). Brand versus brand. *International Commerce Review, 7*(1), 47–53.
Kung, H., & Schmid, L. (2015). Innovation, growth, and asset prices. *The Journal of Finance, 70*(3), 1001–1037.
Lamb, C. W., Jr., Hair, J. F., Jr., & McDaniel, C. (1996). *Marketing* (3rd ed.). Cincinnati, OH: South-Western.
Lee, J. (1997, April 24). Sainsbury ends copycat battle. *Marketing*, London.
Lindsay, M. (2004). Editorial: Achieving profitable growth through more effective new product launches. *Journal of Brand Management, 12*(1), 4–10.
Lundvall, B. A. (1992). *National systems of innovation: An analytical framework*. London: Pinter.
Major, M., & McTaggart, J. (2005, November 15). Blueprints for change. *Progressive Grocer*, pp. 89–94.
McAdam, R. (2005). A multi-level theory of innovation implementation: Normative evaluation, legitimisation and conflict. *European Journal of Innovation Management, 8*(3), 373–388.
Merrilees, B. (2003). Strong brands and innovation: Paradox resolved. Professorial Lecture, 6 November, Griffith University, Australia.
Merrilees, B., & Miller, D. (2001a). Innovation and strategy in the Australian supermarket industry. *Journal of Food Products Marketing, 7*(4), 3–18.
Merrilees, B., & Miller, D. (2001b). Radical service innovations: Success factors in retail pharmacies. *International Journal of New Product Development and Management, 3*(1), 45–58.
Mintzberg, H. (1983). *Power in and around organisations*. Englewood Cliffs, NJ: Prentice Hall.
Naumann, E., & Reck, R. (1982). A buyer's bases of power. *Journal of Purchasing and Materials Management, 18*(4), 8–14.
Nelson, W. (2002). All power to the consumer? Complexity and choice in consumers' lives. *Journal of Consumer Behaviour, 2*(2), 185–195.
Nielsen. (1992). *Category management: Positioning your organisation to win*. Chicago, IL: NTC Business Books.

Nohria, N., & Gulati, R. (1996). Is slack good or bad for innovation? *Academy of Management Journal, 39*(5), 1245–1264.
Ogbonna, E., & Wilkinson, B. (1998). Power relations in the UK grocery supply chain: Developments in the 1990s. *Journal of Retailing and Consumer Services, 5*(2), 77–86.
Olbrich, R., Hundt, M., & Jansen, H. C. (2016). Proliferation of private labels in food retailing: A literature overview. *International Journal of Marketing Studies, 8*(8), 63–76.
Panigyrakis, G., & Veloutsou, C. A. (2000). Problems and future of the brand management structure in the fast moving consumer goods industry: The viewpoint of brand managers in Greece. *Journal of Marketing Management, 16*(1/3), 165–184.
Patchen, M. (1974). The locus and basis of influence on organisational decisions. *Organisational Behaviour and Human Performance, 11*(2), 195–211.
Pfeffer, J. (1981). *Power in organisations*. Marshfield: Pitman.
Raven, B. H. (1993). The bases of power: Origins and recent developments. *Journal of Social Issues, 49*(4), 227–251.
Raven, B. H., & Kruglaski, A. W. (1970). Conflict and power. In P. Swangle (Ed.), *The structure of conflict* (pp. 69–109). New York, NY: Academic Press.
Robert, M. (1995). *Product innovation strategy: Pure & simple*. New York, NY: McGraw-Hill.
Rudder, A. (2003). An evaluation of the NPD activities of four food manufacturers. *British Food Journal, 105*(7), 460–476.
Rundh, B. (2005). The multi-faceted dimension of packaging: Marketing logistic or marketing tool? *British Food Journal, 107*(9), 670–684.
Schaafsma, S., & Hofstetter, J. (2005). Raising the game to a level. *ECR Journal, 5*(1), 66–69.
Sibley, S. D., & Michie, D. A. (1982). An exploratory investigation of cooperation in a franchise channel. *Journal of Retailing, 58*(4), 23–45.
Stanković, L., & Končar, J. (2014). Effects of development and increasing power of retail chains on the position of consumers in marketing channels. *Ekonomika Preduzeća, 62*(5–6), 305–314.
Stern, L. W., & El-Ansary, A. I. (1977). *Marketing channels*. Englewood Cliffs, NJ: Prentice Hall.
Sutton-Brady, C., Kamvounias, P., & Taylor, T. (2015). A model of supplier–retailer power asymmetry in the Australian retail industry. *Industrial Marketing Management, 51*, 122–130.
Thompson, J. D. (1967). *Organizations in action*. New York, NY: McGraw-Hill.

Toffler, A. (1990). *Powershift: Knowledge, wealth and violence at the edge of the 21st century*. New York, NY: Bantam Books.
Tushman, M. L., & Romanelli, E. (1985). Organisational evolution: A metamorphosis model of convergence and re-orientation. In B. Staw & L. Cummings (Eds.), *Research in organisational behaviour* (Vol. 7, pp. 171–222). Greenwich, CT: JAI Press.
Urbanski, A. (2001). Captains courageous: Category management techniques for grocery industry. *Supermarket Business, 56*(11), S3.
Verhoef, P. C., Nijssen, E. J., & Sloot, L. M. (2000). Strategic reactions of national brand manufacturers towards private labels: An empirical study in The Netherlands. *European Journal of Marketing, 36*(11/12), 1309–1326.
Weitz, B., & Wang, Q. (2004). Vertical relationships in distribution channels: A marketing perspective. *Antitrust Bulletin, 49*(4), 859–876.
Wilkinson, I. F. (1973). Power and influence structures in distribution channels. *European Journal of Marketing, 7*(2), 119–129.
Wilkinson, I. F. (1979). Power and satisfaction in channels of distribution. *Journal of Retailing, 55*(2), 79–94.
Zairi, M. (1995). Moving from continuous to discontinuous innovation in FMCG: A re-engineering perspective. *World Class Design to Manufacture, 2*(5), 32–37.

4

Private Label and Manufacturer Brand Coexistence

Overview

This chapter examines specific issues on the coexistence of private label and manufacturer brands, and focuses on private label portfolio and private label share in relation to manufacturer brands. Topics under consideration include retail concentration and its impact on private label performance, the strategic dependency between private label and manufacturer brands, and the relevance of power to the coexistence of the two types of brands. Specifically, the chapter addresses product categories and brands, private label portfolio, the balancing of private label and manufacturer brands, private label share in environments of high retail concentration, and private label/manufacturer brand category share and equilibrium.

Product Categories and Brands

Product categories are the building blocks of the supermarket store (ACNielsen et al. 2006), and the categories are made up of brands. Product categories are important from a category management

R. Chimhundu, *Marketing Food Brands*,
https://doi.org/10.1007/978-3-319-75832-9_4

perspective, but since the categories are actually driven by brands it is reasonable to assume that the management of these brands is a key factor within product categories. In any category, there would usually be quite a number of competing brands from brand manufacturers, and there would be brands from the store side as well. In terms of supermarket product categories, this means the two types of brands (manufacturer brands and private label), would be featuring.

Manufacturer Brands and Private Label

Brands that belong to manufacturers have been referred to using alternative terms such as manufacturer brands and national brands. In the literature, these terms are used interchangeably. This book makes use of the term "manufacturer brand" as employed by Brassington and Pettitt (1997) and Klaus et al. (2006). As an operational definition for this book, manufacturer brands are products that are owned and branded by manufacturers. Brands belonging to retailers have also been made reference to using alternative terms such as retailer own brands, retailer brands, own brands, store brands, house brands, supermarket own brands, private label, private labels, private label brands, private brands, own-label products, retail brands and distributor brands. These terms are largely used interchangeably in the literature. The term "private label" has achieved wide usage. It is, however, noted by some scholars (e.g. Burt and Davis 1999; de Chernatony 1989) that the actual term "label" is associated with a constrained marketing role, which has its emphasis on the packaging aspect (Veloutsou et al. 2004). According to Fernie and Pierrel (1996), "The degree to which a retailer becomes involved in true branding whereby it challenges the leading manufacturer in product positioning would make a distinction between labelling and branding" (Fernie and Pierrel 1996: 48). Some authors have therefore made use of the terms "retailer brand" or "retailer own brand", such as Ailawadi and Keller (2004), Brassington and Pettitt (1997), Burt (2000) and Klaus et al. (2006). In a number of developed economies, private label has graduated from being a basic cheap-quality product that was only positioned at low prices to something much more than that in terms of branding. The operational definition used in this book, adapted from Schutte (1969), is that private labels are products that are owned and branded by retailers; the term "private label" has been chosen as this term is most commonly used in most markets.

Hierarchy of Participating Brands in the Categories

On the manufacturer brand side, participating brands and firms can be conceptualised in terms of Kotler's (2000) hypothetical market structure, which has four classifications: market leader, market challenger, market follower and market nicher. Although this typology is generally taken to refer to participating firms in a market, it has also been taken to classify participating brands in a category. The market-leading brand is the one with the highest share in the category and would normally be at the forefront of distribution coverage, new product introductions, promotional intensity and price changes. In a hypothetical category of four brands ranging from the largest to the smallest, the second brand would be the challenger, the third would be the follower, and the fourth, the nicher. In reality, though, there are many brands participating in the categories such that it is not uncommon to have a variety of brands in each of these four classifications. Some authors have preferred to use terms such as "market leader [...] number two and three branded suppliers and the smaller niche players" (Dewsnap and Jobber 1999: 388), while others describe the structure as consisting of the market leader, secondary suppliers and small brand suppliers (Hogarth-Scott 1999). These authors are nevertheless in agreement on the general hierarchy of participating brands in a category. In this book, Kotler's (2000) typology is used to describe category participants where applicable. It is also important to note that the private label (just like the manufacturer brand) can actually assume any one of these classifications in the category hierarchy depending on its size and activities. Furthermore, manufacturer brands that participate in the categories can, from a quality perspective, fall into three broad segments: premium, standard (middle level) and economy brands, largely differentiated as quality/price segments, and of course branding aspects have something to do with these segments as well. Private label falls into quality/price tiers and the composition of the tiers offered by the private label in a category constitutes the private label portfolio.

Private Label Portfolio

A private label portfolio is part and parcel of retailer brand strategy, and the portfolio of a retail chain can consist of the full spectrum of three quality–price tiers: Tier 1, economy; Tier 2, standard (or medium); and Tier 3, premium. The portfolio can be described from the viewpoint of generations (Anselmsson and Johansson 2009). A four-generation classification (Laaksonen 1994; Laaksonen and Reynolds 1994) was developed to conceptualise private label development. The classification looks at private label from an evolution perspective, grouping them into first, second, third and fourth generations.

From the point of view of the private label quality spectrum, first-generation private labels are generic, have no name, use simple technology and are of lower quality and image than leading manufacturer brands. Second-generation private labels are of medium quality but are seen as lower than leading manufacturer brands, and lag behind market leaders on technology. Third-generation private labels are of a quality that is comparable to leading manufacturer brands and are close to the leading brands on technology. Fourth-generation private labels are of similar or better quality than leading manufacturer brands. They are innovative and different from the top manufacturer brands, and are also strong on the technology dimension.

The four-generation classification can be matched with the three-tier private label portfolio. In the context of the three quality–price tiers that are commonly used with respect to private label (e.g. Coriolis Research 2002), Tier 1 would be first- and second-generation brands; Tier 2, third-generation brands; and Tier 3, fourth-generation brands. The top/premium private labels are therefore classified as Tier 3/fourth generation. Although the four generations of private label development are conceptualised in terms of evolution, it should be noted that more than one generation can actually exist in a category at the same time. The different generations can be marketed at the same time as a portfolio. In addition, introduction of the tiers can follow any order. For instance, it is noted by Fernie and Pierrel (1996) that "in the UK the second-generation brands were in existence prior to the launch of generics in the 1970s" (p. 49). A key factor, however, that is relevant to this book and is worth noting, is the fact that value-added brands in the fourth generation are more innovative and are modelled along the lines of premium quality manufacturer brands. The four-generation and three-tier classifications are appropriate to use in examining private label programmes in grocery retail categories. Such examination would include establishing the kind of quality spectrum that is employed in a particular category, and why.

Furthermore, it is important to note that these quality–price tiers are not just about the intrinsic quality of the brand, but are also largely to do with its image, which is achieved through branding, advertising and related marketing activities. So, other than functional performance, the emotional aspects also have a huge stake in the quality/price tiers. Two FMCG brands that have the same intrinsic quality and deliver the same functional performance may still be perceived differently by consumers, with one being seen as superior to the other, and it is usually marketing activities around the brand that can make the difference. Therefore, brand/category support activities are important in this respect. Logically, enhancing the functional and emotional aspects associated with any brand, whether manufacturer brand or private label, calls for resources (e.g. financial support or R&D facilities) and expertise (knowledge and skill) on the part of the owners/marketers of the brand. With respect to manufacturer brands and private label, whoever has got the resources and expertise is expected to do a good job in this area.

Additionally, one may further reason that, while having the resources is necessary, that in itself has to be complemented by willingness to put the resources behind the brands on the part of the brand owners. According to Howe (1990), in the product categories, manufacturer brands have developed countervailing power by developing their brand franchises. It is noted in the literature, on the other hand, that some private label brands have also significantly improved their quality and brand franchise (ACNielsen et al. 2006; Buck 1993; Burt 2000; Coriolis Research 2002), and "Retailers are producing better private label products, and thereby, strengthening their bond with the consumer" (ACNielsen et al. 2006: 337). Nevertheless, some researchers (e.g. Pauwels and Srinivasan 2004) have argued that private label entry into a category rarely results in the expansion of the category. Manufacturers have also claimed that private labels are not capable of growing the market but simply steal share away from manufacturer brands. Some studies, however (e.g. Putsis and Dhar 1996), have demonstrated that private label is capable of expanding category expenditure. The position this author takes is that grocery retail chains may choose to improve product quality and brand franchise or not to, as a result of rational business judgement on their part. Similarly, they may choose to put more resources into advertising and promotion or not to. Retailer strategic choice therefore determines the resultant private label strategy with reference to aspects such as quality, brand franchise, brand portfolio and promotional issues. Therefore, it is envisaged that retailer strategic thinking with respect to private label is relevant to the study of the coexistence of manufacturer brands and private label in FMCG/supermarket categories, as it has the potential to shape the nature of the coexistence. Establishing such thinking can be better achieved through discussions with the retailers themselves.

Originally, private label brands were marketed as economy brands. It is noted (Kumar and Steenkamp 2007b: 48) that the brands offered higher margins for retailers and allowed differentiation from other retailers; however, "at a certain level of private label penetration (around 20 per cent), 'more became less'". The situation would be different, though, if additional volumes came from premium lines of private label brands that would help to increase overall penetration levels (Kumar and Steenkamp 2007b). Research has shown that, contrary to the traditional view, private label consumers are quality-sensitive. Quality is of equal or even greater importance than price in influencing private label purchase (Sethuraman 2006), as evidenced by a number of studies (e.g. Dhar and Hoch 1997; Erdem et al. 2004; Hoch and Banerji 1993; Richardson et al. 1996a, b; Sethuraman 1992). The quality spectrum of any private label portfolio therefore has relevance for private label and manufacturer brand share trends in grocery product categories. Logically, one might reason that a fully developed private label portfolio with the full quality spectrum would put the private label in a better position to pose a greater competitive threat to the manufacturer brand. Additionally, since private label portfolio issues would be a strategic decision area for the retailer, it can be argued that retailer strategic thinking on the participation of the private label brand in the category has relevance for the way in which private label and manufacturer brands would be balanced in the categories.

Balancing Private Label and Manufacturer Brands in the Product Categories

Key parties to the grocery retail category set-up, that is the grocery retailers, FMCG manufacturers and consumers, are the players whose expectations, actions and interactions (Hoch and Banerji 1993) contribute to the determination of the composition of manufacturer brands and private label in FMCG/supermarket product categories, and each has some sort of power in this area. Consumers have the power of choice. They have the power to decide what to buy and what not to from the competing offerings. Consumers can also be protected by certain laws and regulations. Manufacturers bring the much-needed manufacturer brands to the stores and have brand franchise. Retailers own the distribution channels and also own private label brands that compete with manufacturer brands in the categories. Retailers have increasing power and dominance as a result of consolidation and concentration, increased utilisation of information technology, the growth of private label brands, retailer marketing approach (Hogarth-Scott 1999; Nielsen 1992; Sullivan and Adcock 2002) and category management operational practices; it can therefore be argued that, despite the expectations and actions of consumers and manufacturers, retailers have the capacity to significantly influence the final composition of manufacturer brands and private label in the categories.

In category rationalisation, for instance, even though the consumer has the power of choice and consumer data are taken into account, grocery retailers tend to have the final say on how many and which competing brands to stock. Hypothetically, out of, say, six manufacturer brands (U, V, W, X, Y and Z) in a category, if the category rationalisation process reduces the number of manufacturer brands from six to three and brands W, Y and Z get axed from the category for reasons such as duplication or conflict between manufacturer and retailer, the consumer is forced to choose from the remaining brands. The consumer may not necessarily switch to another retailer, and if they do, the other retailer may be rationalising as well. So, the consumer's so-called power of choice can be restricted to a few brands. The consumer will be "forced" to choose from what is in stock and eventually come to accept it. Again, from a category management perspective, although there may be systematic mechanics that are followed with respect to stocking, shelf-space and position allocation, category rationalisation, product deletion and so on, it can be argued that retailer strategic objectives take precedence over all other determinants, provided that the powerful retailers will benefit. In addition, grocery retailers are the legitimate owners of the supermarket shelves and grocery retailing is their core business; so they are expected to have the final say on category matters.

Furthermore, it can be reasoned that in the coexistence of manufacturer brands and private label, a retailer strategic objective to raise private label share to a certain level in a category may mean tilting certain merchandising measures in favour of private label. And the converse can be true as well; the retailer may want to ensure that the share of their own private label brand does not go beyond a certain level in a category, for strategic reasons such as promoting long-term equilibrium points. Such strategic category management regimes are expected to be driven by the retailer because the balance of power in the equation is in favour of the retailer. These aspects, therefore, are worth investigating in depth in this book. How these two types of brands are balanced, particularly in an FMCG landscape characterised by high retail consolidation and concentration, is examined. In addition, an investigation of the power bases that are dominant in this context is a related area that would serve to theoretically contextualise the discussion.

Private Label Share in an Environment of High Retail Concentration

Retail consolidation is measured by use of a concentration ratio which is the percentage of sales commanded by the largest retail firms in an industry (Defra 2006). Retail consolidation gives more power to grocery retail chains in relation to manufacturers because of economies of scale and scope (Cotterill 1997) and related factors such as centralised buying and distribution. Grocery retail consolidation/concentration has been linked with pushing private label share to high levels (Burt 2000; Coriolis Research 2002; Cotterill 1997; Defra 2006; Hollingsworth 2004; Nielsen 2014; Rizkallah and Miller 2015). Theories have been advanced (Galbraith 1952; Porter 1976) that an increase in the relative power of grocery retailers in relation to suppliers would cause a rise in the aggregate market share of grocery private label brands. Similarly, theory drawn from the experiences of a number of countries suggests that retail consolidation is a key driver of private label growth (Coriolis Research 2002). In addition, ACNielsen (2005) found that, of the top ten most developed private label countries in the world, nine have high retail concentrations of above 60%, where retail concentration is measured as the proportion held by the top five retailers in each country (i.e. the five-firm concentration ratio). Retail consolidation and concentration are therefore relevant factors in private label/manufacturer brand dynamics.

Private Label/Manufacturer Brand Category Share and Equilibrium

Central to the topic of the coexistence of manufacturer brands and private label in FMCG/supermarket product categories are aspects related to private label/manufacturer brand share (and trends), the balancing of the two types of brands, shelf management, and related competition issues between the two types of brands. This book seeks to better understand the hows and whys of strategic issues related to the coexistence of manufacturer brands and private label in the product categories.

Previous research has investigated the topic from a number of perspectives. Most studies have tended to look at the link between private label share trends and business cycles/economic factors or industry factors (e.g. Baden-Fuller 1984; Coriolis Research 2002; Hoch and Banerji 1993; Hoch et al. 2002a, b; Lamey et al. 2005, 2007; Quelch and Harding 1996). Some of the research has investigated the subject from a historical perspective (e.g. Hernstein and Gamliel 2004; Steiner 2004). Other works have dwelt on competition issues related to the coexistence of the two types of brands (e.g. Cotterill et al. 2000; Hultman et al. 2008; Mills 1999; Miranda and Joshi 2003; Verhoef et al. 2002); and some of the research work has investigated the subject from a merchandising perspective (e.g. Gomez and Rubio 2008; Saurez 2005).

Baden-Fuller (1984) found that while industry factors are important in influencing private label, the strategies of the companies themselves are actually more important. Hoch and Banerji (1993) observed that private label share tends to be higher in categories that have higher dollar sales, higher gross profit margin, fewer manufacturer brands and less manufacturer brand advertising spending.

ACNielsen (2005) and Coriolis Research (2002) have demonstrated that retail consolidation and concentration can lead to greater private label penetration. Similarly, Lamey et al. (2005, 2007) confirmed the link between the success of private label and economic expansions and contractions, and also found that "consumers switch more extensively to store brands during bad economic times than they switch back to national brands in a subsequent recovery" (2007, p. 1). These studies show that private label share trends and the nature of coexistence between manufacturer brands and private label (in this situation, measured by share of a respective category) can be influenced by economic and industry factors, but as observed by Baden-Fuller (1984), the strategies of the respective firms themselves are important.

With respect to influence by the strategies of the firms, this author puts forward the suggestion that these can be looked at from the viewpoint of the private label strategies of grocery retail chains and the counterstrategies of manufacturers/suppliers, as well as the strategies of manufacturers/suppliers and the respective counterstrategies of grocery retail chains. While traditionally manufacturer brands have tended to dominate FMCG categories, the increasing power of grocery retail chains (owners and managers of private label) and their emphasis on private label brands and the category management practice make the examination of retail chain strategies with regard to retailer brands even more important. Most research, however, has tended to focus on economic and industry factors at the expense of strategy factors; specifically, grocery retail chain strategy with respect to how the manufacturer brand and the private label should coexist, and why, has not really been subjected to intense investigation.

A study by Hoch et al. (2002a, b) showed the crucial role of strategy factors in influencing private label, thereby giving support to the suggestion of Baden-Fuller (1984) on the role played by private label strategy. Hoch et al. (2002a, b) investigated 225 consumer goods categories (consisting of both manufacturer brands and private label) for an eight-year period from 1987 to 1994 in the USA and found that the private label brand showed positive market share evolution as compared to the manufacturer brand. One explanation was that the private label is the only brand that controls both its own marketing mix decisions and, to a large extent, the marketing mix decisions of its competitors (Hoch et al. 2002a, b). While the study employed robust methodology, a limitation of the study was that it covered a relatively short period of time: eight years. This may tend to give the impression that the same trend would hold outside this eight-year period, which may not be the case.

If the study had looked at more long-term trends, say 15 to 20 years, the situation may have been different. Market share evolution can intensify; it can stabilise or even decline depending on the environment and private label strategy. Again, if the trends were analysed by category, different categories would most likely present different situations as industry circumstances and strategic policies may differ depending on the category. The study also looked at data from one country, and therefore the results would not necessarily hold across all FMCG industries in different economies.

One might expect that in an environment that was more highly concentrated than the USA, where the study was carried out, a more radical market share evolution would be reflected if retailers were to pursue private label strategies that took advantage of their power to influence things in their favour. However, if the retailers were not willing or able to take such advantage, then a different situation would prevail. By extension, a study that assumed that retailers were willing to fully apply their power would in all likelihood pursue a line of investigation with an expectation of increased private label dominance or even overdominance in the coexistence of the two types of brands in the categories. At the same time, a study that did not assume that retailers would be willing to exploit their power to the full, but would rather treat the other party (e.g. manufacturer/supplier) as a business partner who should be looked after, would fittingly have converse propositions. Research based on the latter expectations might find a private label strategy that was working towards an optimum balance between manufacturer brands and its own brands in the categories. Such a balance would provide equilibrium points (between manufacturer and private label) that would be perceived to safeguard the long-term strategic health of the categories.

With regard to research that has investigated the subject from a merchandising perspective, Saurez (2005) showed that there is a positive direct relationship between private label brand shelf space and market share. By implication therefore, if retail chains can use their muscle to tilt merchandising measures in favour of their private label, this would be expected to act in the retail chains' favour from a market share perspective. Whether they do this or not, and why, is part of the investigation agenda for this book. In an earlier section of the book, Chap. 3, consumer choice section, private label has been described as a sacred cow in the category management set-up (Major and McTaggart 2005), which means that it can be exempt from the normal category management rules and receive favourable treatment if need be. However, studies have produced conflicting results on this subject. For instance, Bell and Duder (1998) showed that grocery retailers were not favouring their private label brands with respect to shelf facings and better position on the shelves. Conversely, a much later study by Gomez and Rubio (2008) found that "on average, manufacturers consider that retailers are favouring unequal competition terms between manufacturer and store brands" (p. 50). Current anecdotal evidence has supported this latest research result. The conflict in these studies could be explained by the fact that, as far back as 1996, the year the data for Bell and Duder (1998) were collected in New Zealand, grocery retail consolidation and concentration in the country had not reached the level it is at today, and nor had the emphasis on private label development and category management. Therefore, the current situation may be expected to be consistent with Gomez and Rubio's (2008) study, although it was carried out in a different environment, Spain. Or the explanation may simply be attributed to differences in private label strategies by the respective retail chains involved in the different FMCG industries. Even then, one might still reason that private label brand strategies that were suitable more than a decade ago may not really be justifiable in the radically altered FMCG landscape (Kumar and Steenkamp 2007a) characterised by increased retailer power and direct competition between private label and manufacturer brands. These issues therefore justify the investigation of strategic management regimes governing the coexistence of manufacturer brands and private label in the categories. Such regimes may not be standard in the industry set-ups of different economies, and a focused, intensive study of a specific FMCG/supermarket industry would yield valuable insights into the hows and whys of manufacturer brand/private label coexistence in grocery retail categories.

With respect to the research stream that has looked into coexistence from the viewpoint of competition between the two types of brands (e.g. Cotterill et al. 2000; Hultman et al. 2008; Mills 1999; Miranda and Joshi 2003; Verhoef et al. 2002), some of the main themes that have arisen include the use of product innovations and advertising/brand building in the competitive coexistence of manufacturer brands and private label in the categories. The importance of quality in making the private label brand more competitive is also emphasised. However, how these activities influence private label strategic thinking with respect to the coexistence of the two types of brands is an area that needs further investigation.

A snapshot of some of the relevant academic research (most of which has already been discussed) on the coexistence of manufacturer brands and private label is presented in Table 4.1.

Table 4.1 Key Literature on manufacturer brands and private label

Author(s) (year)	Type of study/ method	Environment	Findings/summary statement on aspects related to the study on manufacturer and retailer brands
Baden-Fuller (1984)	Quantitative: statistical correlation	UK	Strategies of retailers important in determining retailer brand share, not just industry factors. To fully assess what is likely to happen in the future, research needs to understand the area of "retail strategic thinking in the area of retail brands" (p. 525).
Hoch and Banerji (1993)	Quantitative: regression	USA	Retailer brands do better in the larger categories offering high margins, or when competing with fewer manufacturer brands that spend less on national advertising. High retailer brand quality is of greater importance than lower price. Low retailer brand share in a category "does not imply that a particular retailer cannot create a successful program in that category" (p. 66).
Mills (1999)	Quantitative: modelling	USA	The most effective counterstrategies of manufacturer brands against private label brands involve, among other things, having "a technological comparative advantage (widening the quality gap)" (p. 143).

(*continued*)

Table 4.1 (continued)

Author(s) (year)	Type of study/ method	Environment	Findings/summary statement on aspects related to the study on manufacturer and retailer brands
Cotterill et al. (2000)	Quantitative: linear approximate almost ideal system	USA	For either the manufacturer brand or the private label, a price increase lowers market share. Research involving competitive interaction between manufacturer brands and private label encouraged.
Hoch et al. (2002a, b)	Quantitative: regression, non-parametric tests, proportional draw analysis	USA	On average, private label can grow while manufacturer brand exhibits no growth because, in addition to the private label controlling its own marketing spending, it does exert influence over the marketplace spending of manufacturer brand competitors. Continued "empirical and theoretical research into the unique behaviour of private labels" (p. 27) encouraged.
Verhoef et al. (2002)	Quantitative: factor analysis, cluster analysis	Netherlands	Manufacturer brand competition with private label is less direct and rather subtler. Focus is on "increasing the distance from private labels, thereby using advertising [for brand strength] or product innovations [technological innovations]" (p.1323), strategies that are considered to be effective. Future research on the strategy aspects of manufacturer brands and private label encouraged.

(continued)

Table 4.1 (continued)

Author(s) (year)	Type of study/ method	Environment	Findings/summary statement on aspects related to the study on manufacturer and retailer brands
Miranda and Joshi (2003)	Quantitative: logit log linear analysis procedure	Australia	Quality of private label generally more appealing to the consumer than price. To be competitive with private label, investment in private label programmes encouraged.
Herstein and Gamliel (2004)	Qualitative: historical	Mainly North America and Europe (but has a global ingredient)	Five eras identified in the history of private labels. Private label penetration and development higher in Europe and North America than in the developing world.
Steiner (2004)	Qualitative: historical	N/A	Private label brands of large grocery retail chains have unique competitive weapons that can limit the market power of powerful manufacturer brands; these weapons are not possessed by rival manufacturer brands. In some categories, depending on category management arrangements, competition can be replaced by collusion, engineered by the CC.
Saurez (2005)	Quantitative: neural network analysis	Spain	There is a direct relationship between private label shelf space and market share. Moreover, overmerchandising the private label brand may damage overall category profitability.

(*continued*)

Table 4.1 (continued)

Author(s) (year)	Type of study/ method	Environment	Findings/summary statement on aspects related to the study on manufacturer and retailer brands
Sethuraman (2006)	Review	Largely USA	Common management beliefs on the best way to market private label are assessed and either supported, negated or refined. Further, "More research needed [...] on private-label marketing in other parts of the world" (p. 41) than in the USA. Further research needed on "the relationship between channel power and private label share" (p. 41).
Kumar and Steenkamp (2007a, b)	Review	USA, UK and Europe	Private label brands are no longer cheap versions of manufacturer brands. Innovation on the part of manufacturer brands is important in rising to the private label challenge.
Lamey et al. (2007)	Quantitative: time series/ econometric	USA, UK, Belgium and Germany	Private label share is positively linked to business cycles, that is economic expansions and contractions. However, "consumers switch more extensively to store brands during bad economic times than they switch back to national brands in a subsequent recovery" (p. 1).
Hultman et al. (2008)	Qualitative: multiple case study of four manufacturing companies	Sweden	Private labels have the advantage of "overall control of the market in which they operate", and manufacturer brands have the advantages of "product development and superior brand reputation" (p. 125).

(*continued*)

Table 4.1 (continued)

Author(s) (year)	Type of study/ method	Environment	Findings/summary statement on aspects related to the study on manufacturer and retailer brands
Gomez and Rubio (2008)	Quantitative: multivariate analysis	Spain	Manufacturers consider that retailers favour own brands through better merchandising. However, different groups of manufacturers have different perceptions.
Kremer and Viot (2012)	Qualitative and quantitative	France	Store brands have a positive impact on the image of the retailer.
Leingpibul et al. (2013)	Quantitative: structural equation modelling	USA	Customer purchase behaviour is influenced by brand loyalty, which is greater for private label than for manufacturer brands.
Sethuraman and Gielens (2014)	Meta-analysis	USA	Study shows that as many as 20 determinants of store brand share have empirically generalisable effects.

Source: Created for this book based on the literature

It should be noted that this table largely focuses on refereed journal articles, and that it is not meant to be exhaustive. It is rather designed to be comprehensive enough to incorporate major research and issues that are relevant to this book. The following points can be deduced from Table 4.1 with regard to opportunities for further research that would make a contribution to the literature on the coexistence of manufacturer brands and private label in the FMCG/supermarket product categories.

Firstly, there is, in combination, a strong collective call for further research on manufacturer brands and private label as outlined in the "Findings" column of the table: Sethuraman (2006) has recommended further research on "the relationship between channel power and private label share" (p. 41); Baden-Fuller (1984) has called for further research on "retail strategic thinking in the area of retail brands" (p. 525); Verhoef et al. (2002) have encouraged more research on the strategy aspects of manufacturer brands and private label; Hoch et al. (2002a, b) have recommended further research on "the unique behaviour of private labels" (p. 27); Cotterill et al. (2000) have encouraged more research on the competitive interaction between manufacturer brands and private label brands; and Herstein and Gamliel (2004) have called for more research on the meaning and importance of private label from the point of view of the retailers, manufacturers and consumers. These recommended areas for further research are very much interlinked. For instance, the powerful retail chains' strategic thinking with respect to private label brands in relation to manufacturer brands is likely to influence how they employ their power in dealing with coexistence issues, such as their approach to gaining private label share and to merchandising. It could also influence whether they use their power in a predominantly coercive or non-coercive way, or a combination of both.

It is this author's position that the strategic thinking of the retailer would take precedence over the strategic thinking of the manufacturer on private label/manufacturer brand coexistence issues because of the balance of power, which is in favour of the retailer. Such coexistence issues include, inter alia: strategic objectives regarding the level to which the private label should grow in the categories in terms of share (in relation to the manufacturer brand); strategic decisions concerning merchandising measures between the two types of brands; competitive strategy aspects of private label in relation to manufacturer brands; how areas of strategic dependency between the two types of brands are handled and how much value is attached to such strategic dependency, as well as whether recognition for such strategic dependency would be factored into the determination of strategic management regimes governing the coexistence of the two types of brands in the categories.

Secondly, a glance at the table shows that most of the studies were carried out in the USA, the UK and Europe. Countries such as New Zealand have not served as research environments for much of the published academic research in the area. However, private label research experts have identified the need to take private label research to other environments than those that are frequently researched: "More research [is] needed [...] on private-label marketing in other parts of the world" (Sethuraman 2006: 41). Taking the research to countries such as New Zealand and

Australia, among others, would contribute to the academic literature in a way that would offer a fresh perspective, as these under-researched environments on the subject offer different conditions in comparison to the USA and Europe. Even by 2017, mainstream research publications on private label and manufacturer brands were not generally from the "other parts of the world" referred to.

In New Zealand, it is largely commercial research that has been undertaken on private label and manufacturer brands. The most notable are Coriolis Research (2002) and subsequent Coriolis Research reports, as well as ACNielsen data. These have been cited in earlier discussions in this book. Such research is extremely useful for business decision making, and it still contributes to the body of knowledge, however it is not academic research as such. Some of the research of academic nature (though not published in academic journals) that has been carried out on manufacturer brands and private label, such as Bell and Duder's (1998) study cited earlier, is rather dated as environmental factors have changed. Furthermore, Keen (2003) investigated only the dairy sector with respect to private label development, and the research is rather limited in scope although it identified relevant issues such as the contributions of manufacturer brands in the areas of market development and innovation. Additionally, Chimhundu (2004) and Chimhundu and Hamlin (2007, 2008) incorporated the issue of manufacturer brand and private label into a larger study that was dealing with brand management. Because of the limitation of treating the subject only as a minor facet of a study handling a number of issues, the scope of the research did not allow adequate, in-depth investigation. As a result, the research can only be credited for raising relevant issues that would need to be investigated in a larger study dedicated specifically to the coexistence of manufacturer brands and private label in FMCG/supermarket product categories. These include, among others, the issue of innovation and brand development in the coexistence of manufacturer brands and private label (e.g. Chimhundu 2004; Chimhundu and Hamlin 2007, 2008; Keen 2003). This is in addition to the contradiction in the literature regarding, on the one hand, an adversarial private label approach to gaining penetration/share and growth, and on the other, a less aggressive and more accommodating stance that seeks to harness manufacturer brands' contributions (Chimhundu et al. 2011).

Thirdly, from a methodological perspective, it can be noted from the table that most of the research has employed the quantitative methodology. Qualitative empirical studies that allow in-depth analysis of specific issues in the coexistence of manufacturer brands and private label have been scarce. From a category management perspective in this respect, Lindblom and Olkkonen (2006) have bemoaned the lack of qualitative, empirical studies. For this reason, this book takes the opportunity to focus on the hows and whys of manufacturer brand and private label coexistence.

Research Direction of this Book

The review of the literature in this chapter and in the previous two chapters has discussed a number of aspects related to the coexistence of manufacturer brands and private label in grocery retail product categories. These include the category management set-up itself, product innovation, category support and consumer choice, retail consolidation/concentration and the theory of power, as well as specific coexistence issues related to manufacturer brands and private label. The focus of this book is on how the two types of brands coexist in a radically shaped FMCG landscape characterised by high retail consolidation and concentration and increased retailer power, and why? The study employs a primary research question that is further decomposed into subsidiary questions that then serve as a guide in the development of research issues. Therefore, the primary research question is: *How do manufacturer brands and private label coexist in FMCG/supermarket product categories in a grocery retail landscape characterised by high retail concentration, and how relevant is power to this coexistence?*

The primary research question is further decomposed into three subsidiary research questions. The literature review has discussed the link between retail consolidation/concentration, retailer power and private label penetration/share. It is expected that the higher the concentration level, the greater the retailer power in relation to manufacturers and the greater the likelihood of private label category dominance or overdominance of the FMCG/supermarket categories. It is notable, though, that in the categories, manufacturer brands also collectively hold a considerable level of countervailing power. It is therefore important to test the connection between retail concentration and private label share in environments with different levels of retail concentration, with the main focus being on countries with the highest level of retail concentration such as New Zealand, which is a duopoly. Therefore, the first subsidiary research question is: *Does a grocery retail environment characterised by high retail concentration lead to an overdominance of private label in relation to manufacturer brands in FMCG/supermarket product categories?*

The literature review has also discussed manufacturer brand and private label innovation and category support activities as an area of strategic dependency between the two types of brands. These aspects are seen as having the potential to exert an influence on the determination of strategic policies that govern the coexistence of the two types of brands in the categories. This is a perspective that has not featured much in the mainstream academic literature. Consumer focus, as emphasised by the category management practice, is the factor that is most commonly recognised as having an influence on such strategic policies related to the composition of manufacturer brands and private label in the categories. Thus, the second subsidiary research question is: *How important are aspects of strategic dependency between manufacturer brands and private label in determining the nature of coexistence between the two types of brands in FMCG/supermarket product categories?*

Furthermore, the review of the literature has discussed the theory of power in relation to the coexistence of manufacturer brands and private label, and particularly in relation to a grocery retail environment characterised by high retail concentration and a balance of power that is largely in favour of the retailers. The role played by power here is important to assess, as are the dominant bases of power, as they are a reflection of retailer strategic thinking. This area forms the basis for the third subsidiary research question, which is: *In an FMCG/supermarket landscape characterised by high retail concentration and direct competition between brands owned and managed by owners of the grocery retail shelves (private label) and those owned and managed by their suppliers (manufacturer brands), what is the role of power in the coexistence relationship between the two types of brands in the product categories?*

These research sub-questions will be further decomposed into specific research issues in the next chapter. Addressing the research issues and answering the questions will serve to advance the literature by way of adding "smaller bricks of new knowledge" (Lindgreen et al. 2001: 513) in the area of the coexistence of manufacturer brands and private label in FMCG/supermarket product categories. This would mark an "incremental step in understanding" (Phillips and Pugh 2000: 64) the coexistence of manufacturer brands and private label in FMCG/supermarket product categories in a radically altered (Kumar and Steenkamp 2007a) FMCG landscape characterised by high retail consolidation/concentration as well as direct competition between private label and manufacturer brands.

Chapter Recap

This chapter has explored product categories and brands, private label portfolio, balancing manufacturer brands and private label in the product categories, private label share in an environment of high retail consolidation/concentration, private label/manufacturer brand category share and equilibrium, and the research direction for this book. The next chapter covers key research issues in the marketing of private label and manufacturer brands.

References

ACNielsen. (2005). *The power of private label: A review of growth trends around the world*. New York, NY: ACNielsen.

ACNielsen, Karolefski, J., & Heller, A. (2006). *Consumer-centric category management: How to increase profits by managing categories based on consumer needs*. Hoboken, NJ: Wiley.

Ailawadi, K. L., & Keller, K. L. (2004). Understanding retail branding: Conceptual insights and research priorities. *Journal of Retailing, 80*(4), 331–342.

Anselmsson, J., & Johansson, U. (2009). Third generation retailer brands: Retailer expectations and consumer response. *British Food Journal, 111*(7), 717–734.

Baden-Fuller, C. W. F. (1984). The changing market share of retail brands in the UK grocery trade 1960–1980. In C. W. F. Baden-Fuller (Ed.), *The economics of distribution* (pp. 513–526). Milan: Franco Angeli.

Bell, J. D., & Duder, J. N. (1998, December). *The battle for shelf-space in New Zealand supermarkets: Do supermarkets favour their own brands with extra and better positioned facings?* Paper presented at the ANZMAC Conference, University of Otago, Dunedin, New Zealand.

Brassington, F., & Pettitt, S. (1997). *Principles of marketing*. London: Pitman Publishing.

Buck, S. (1993). Own label and branded goods in FMCG markets: An assessment of the facts, the trends and the future. *Journal of Brand Management, 1*(1), 14–21.

Burt, S. (2000). The strategic role of retail brands in British grocery retailing. *European Journal of Marketing, 34*(8), 875–890.

Burt, S., & Davis, S. (1999). Follow my leader? Lookalike retailer brands in non-manufacturer-dominated product markets in the UK. *The International Review of Retail, Distribution and Consumer Research, 9*(2), 163–185.

Chimhundu, R. (2004). *Future of the brand management structure in FMCG: A two-dimensional perspective*. MCom thesis, University of Otago, Dunedin, New Zealand, 272 pp.

Chimhundu, R., & Hamlin, R. P. (2007). Future of the brand management structure in FMCG. *The Journal of Brand Management, 14*(3), 232–239.

Chimhundu, R., & Hamlin, R. P. (2008). *The brand management structure in consumer packaged goods: A research monograph on its current status and future prospects*. Saarbrucken, Germany: VDM Publishing.

Chimhundu, R., Hamlin, R. P., & McNeill, L. (2011). Retailer brand share statistics in four developed economies from 1992 to 2005: Some observations and implications. *British Food Journal, 113*(3), 391–403.

Coriolis Research. (2002). *Responding to private label in New Zealand*. Auckland: Coriolis Research.

Cotterill, R. W. (1997). The food distribution system of the future: Convergence towards the US or UK model? *Agribusiness, 3*(2), 123–135.

Cotterill, R. W., Putsis, W. P., Jr., & Dhar, R. (2000). Assessing the competitive interaction between private labels and national brands. *The Journal of Business*, 73(1), 109-137.

de Chernatony, L. (1989). Branding in the era of retailer dominance. *International Journal of Advertising, 8*(3), 245–260.

Defra (Department for Environment, Food and Rural Affairs). (2006). *Economic note on UK grocery retailing*. London, UK: Department for Environment, Food and Rural Affairs, Food and Drinks Economics Branch.

Dewsnap, B., & Jobber, D. (1999). Category management: A vehicle for integration between sales and marketing. *Journal of Brand Management, 6*(6), 380–392.

Dhar, S. K., & Hoch, S. J. (1997). Why store brand penetration varies by retailer. *Marketing Science, 16*(3), 208–227.

Erdem, T., Zhao, Y., & Valenzuela, A. (2004). Performance of store brands: A cross-country analysis of consumer store-brand preferences, perceptions, and risk. *Journal of Marketing Research, 41*(1), 86–100.

Fernie, J., & Pierrel, F. R. A. (1996). Own branding in UK and French grocery markets. *Journal of Product and Brand Management, 5*(3), 48–59.

Galbraith, J. K. (1952). *American capitalism: The concept of countervailing power*. Boston, MA: Houghton Mifflin Company.

Gomez, M., & Rubio, N. (2008). Shelf management of store brands: Analysis of manufacturers' perceptions. *International Journal of Retail and Distribution Management, 36*(1), 50–70.

Herstein, R., & Gamliel, E. (2004). An investigation of private branding as a global phenomenon. *Journal of Euromarketing, 13*(4), 59–77.

Hoch, S. J., & Banerji, S. (1993). When do private labels succeed? *Sloan Management Review, 34*(4), 57–67.

Hoch, S. J., Montgomery, A. L., & Park, Y. H. (2002a). Why private labels show long-term market evolution. *Marketing Department Working Paper*, Wharton School, University of Pennsylvania, PA.

Hoch, S. J., Montgomery, A. L., & Park, Y. H. (2002b). Long-term growth trends in private label market shares. *Marketing Department Working Paper #00-010*. Wharton School, University of Pennsylvania, PA.

Hogarth-Scott, S. (1999). Retailer-supplier partnerships: Hostages to fortune or the way forward for the millennium? *British Food Journal, 101*(9), 668–682.

Hollingsworth, A. (2004). Increasing retail concentration: Evidence from the UK food sector. *British Food Journal, 106*(8), 629–638.

Howe, W. S. (1990). UK retailer vertical power, market competition and consumer welfare. *International Journal of Retail and Distribution Management, 18*(2), 16–25.

Hultman, M., Opoku, R. A., Salehi-Shangari, E., Oghazi, P., & Bui, Q. T. (2008). Private label competition: The perspective of Swedish branded goods manufacturers. *Management Research News, 31*(2), 125–141.

Keen, E. (2003). *Private label development*. BCom Honours dissertation, University of Otago, Dunedin, New Zealand.

Klaus, G., Esbjerg, L., Bech-Larsen, T., Brunsø, K., & Juhl, H. J. (2006). Consumer preferences for retailer brand architectures: Results from a conjoint study. *International Journal of Retail and Distribution Management, 34*(8), 597–608.

Kotler, P. (2000). *Marketing management* (10th ed.). Upper Saddle River, NJ: Prentice Hall Inc.

Kremer, F., & Viot, C. (2012). How store brands build retailer brand image. *International Journal of Retail & Distribution Management, 40*(7), 528–543.

Kumar, N., & Steenkamp, J. E. M. (2007a). *Private label strategy: How to meet the store brand challenge*. Boston, MA: Harvard Business School Press.

Kumar, N., & Steenkamp, J. E. M. (2007b). Brand versus brand. *International Commerce Review, 7*(1), 47–53.

Laaksonen, H. (1994). *Own brands in food retailing across Europe*. Oxford: Oxford Institute of Retail Management.

Laaksonen, H., & Reynolds, J. (1994). Own brands in food retailing across Europe. *The Journal of Brand Management, 2*(1), 37–46.

Lamey, L., Deleersneyder, B., Dekimpe, M. G., & Steenkamp, J. E. M. (2005). The impact of business-cycle fluctuations on private label share. Series Reference Number ERS-2005-061-MKT (pp. 1–41). Erasmus Research Institute of Management.

Lamey, L., Deleersneyder, B., Dekimpe, M. G., & Steenkamp, J. E. M. (2007). How business cycles contribute to private label success: Evidence from the United States and Europe. *Journal of Marketing, 71*(1), 1–15.

Leingpibul, T., Allen Broyles, S., & Kohli, C. (2013). The comparative influence of manufacturer and retailer brands on customers' purchase behavior. *Journal of Product & Brand Management, 22*(3), 208–217.

Lindblom, A., & Olkkonen, R. (2006). Category management tactics: An analysis of manufacturers' control. *International Journal of Retail and Distribution Management, 34*(6), 482–496.

Lindgreen, A., Vallaster, C., & Vanhamme, J. (2001). Reflections on the PhD process: The experience of three survivors. *The Marketing Review, 1*(4), 505–529.

Major, M., & McTaggart, J. (2005, November 15). Blueprints for change. *Progressive Grocer*, pp. 89–94.

Mills, D. E. (1999). Private labels and manufacturer counterstrategies. *European Review of Agricultural Economics, 26*(2), 125–145.

Miranda, M. J., & Joshi, M. (2003). Australian retailers need to engage with private labels to achieve competitive difference. *Asia Pacific Journal of Marketing and Logistics, 15*(3), 34–47.

Nielsen. (1992). *Category management: Positioning your organisation to win.* Chicago, IL: NTC Business Books.

Nielsen. (2014). *The state of private label around the world.* The Nielsen Company.

Pauwels, R., & Srinivasan, S. (2004). Who benefits from store brand entry? *Marketing Science, 23*(3), 364–390.

Phillips, E. M., & Pugh, D. S. (2000). *How to get a PhD: Handbook for students and their supervisors* (3rd ed.). Buckingham: Open University Press.

Porter, M. E. (1976). *Interbrand choice, strategy and market power.* Harvard, MA: American Marketing Association.

Putsis, W. P., & Dhar, R. (1996). Category expenditure and promotion: Can private labels expand the pie? *Working Paper*, Yale University, New Haven, CT.

Quelch, J. A., & Harding, D. (1996). Brands versus private labels: Fighting to win. *Harvard Business Review, 74*(1), 99–109.

Richardson, P., Jain, A. K., & Dick, A. (1996a). Household store brand proneness: A framework. *Journal of Retailing, 72*(2), 159–185.

Richardson, P., Jain, A. K., & Dick, A. (1996b). The influence of store aesthetics on evaluation of private label brands. *Journal of Product and Brand Management, 5*(1/3), 9–28.

Rizkallah, E. G., & Miller, H. (2015). National versus private-label brands: Dynamics, conceptual framework, and empirical perspective. *Journal of Business & Economics Research, 13*(2), 123–136.

Schutte, T. F. (1969). The semantics of branding. *Journal of Marketing, 33*(2), 5–11.

Sethuraman, R. (1992). Understanding cross-category differences in private label shares of grocery products. *Working Paper No. 92-128*, Marketing Science Institute, Cambridge, MA.

Sethuraman, R. (2006). Private-label marketing strategies in packaged goods: Management beliefs and research insights. *Report No. 06-108*, Marketing Science Institute, Cambridge, MA.

Sethuraman, R., & Gielens, K. (2014). Determinants of store brand share. *Journal of Retailing, 90*(2), 141–153.

Steiner, R. L. (2004). The nature and benefits of national brand/private label competition. *Review of Industrial Organization, 24*(2), 105–127.

Suarez, M. (2005). Shelf space assigned to store and national brands: A neural networks analysis. *International Journal of Retail and Distribution Management, 33*(11/12), 858–879.

Sullivan, M., & Adcock, D. (2002). *Retail marketing*. London: Thomson.

Veloutsou, C., Gioulistanis, E., & Moutinho, L. (2004). Own labels choice criteria and perceived characteristics in Greece and Scotland: Factors influencing the willingness to buy. *Journal of Product and Brand Management, 13*(4), 228–241.

Verhoef, P. C., Nijssen, E. J., & Sloot, L. M. (2002). Strategic reactions of national brand manufacturers towards private labels: An empirical study in The Netherlands. *European Journal of Marketing, 36*(11/12), 1309–1326.

5

Key Research Issues in the Marketing of Private Label and Manufacturer Brands

Overview

This chapter summarises the literature reviewed on private label and manufacturer brands in food product categories in a highly concentrated grocery retail landscape. The summary is directly linked to the research issues of this book. The research issues fall into three major themes, which are: balance between private label and manufacturer brands in the product categories, (research issues 1, 2a, 2b and 2c); innovation, category marketing support and consumer choice (research issues 3a, 3b, 4a, 4b, 4c and 5); and category strategic policies on the coexistence of the two types of brands (research issues 6 and 7). The chapter also recaps the research questions, research themes, specific research issues and the conceptual framework.

A Recap of the Research Questions

As established in Chap. 4, the primary research question is: *How do manufacturer brands and private label coexist in FMCG/supermarket product categories in a grocery retail landscape characterised by high retail concentration, and how relevant is power to this coexistence?*

R. Chimhundu, *Marketing Food Brands*,
https://doi.org/10.1007/978-3-319-75832-9_5

The primary research question has been decomposed into three subsidiary research questions, and these are: Does a grocery retail environment characterised by high retail concentration lead to an overdominance of private label in relation to manufacturer brands in FMCG/supermarket product categories? How important are aspects of strategic dependency between manufacturer brands and private label in determining the nature of coexistence between the two types of brands in FMCG/supermarket product categories? And in an FMCG/supermarket landscape characterised by high retail concentration and direct competition between brands owned and managed by owners of the grocery retail shelves and those owned and managed by their suppliers, what is the role of power in the coexistence relationship between the two types of brands in the product categories? A series of specific research issues in the form of smaller sub-questions is employed to address specific aspects of the above research questions.

Research Themes and Specific Research Issues

Balance Between Manufacturer Brands and Private Label FMCG/Supermarket Product Categories

The FMCG/supermarket environment has generally become more consolidated and concentrated, but the levels of concentration vary from economy to economy. Retail consolidation/concentration has been linked with increased private label share (Burt 2000; Coriolis Research 2002; Cotterill 1997; Defra 2006; Hollingsworth 2004; Nielsen 2014; Rizkallah and Miller 2015), in which case, highly concentrated grocery retail environments are expected to have higher private label shares. In addition, it has been established that FMCG retailers largely hold the balance of power in relation to FMCG manufacturers (Berthon et al. 1997; Hogarth-Scott 1999; Hollingsworth 2004; Hovhannisyan and Bozic 2016; Panigyrakis and Veloutsou 2000; Stanković and Končar 2014; Sutton-Brady et al. 2015; Weitz and Wang 2004) and that power retailers have become the gatekeepers of the supermarket shelves (ACNielsen et al. 2006). This may predispose the retailers to capitalise on their power to push their private label shares to high levels in order to exploit private label higher capacity to generate profit (ACNielsen 2005; Burt 2000; Burt and Sparks 2003; Coriolis Research 2002; Cotterill 1997; Defra 2006; Galbraith 1952; Porter 1976). The systematic push could reasonably be perceived to have no end in sight, leading to an overdominance of private label in the categories.

Two perspectives have come out of this issue and these represent a contradiction in the literature (Chimhundu et al. 2011). One line of reasoning that directly relates to this discussion is that it is expected that in highly concentrated grocery retail environments, there is bound to be an overdominance of private label brands. This line of reasoning is consistent with the adversarial and aggressive approach to private label growth on the part of the retail chains that is discussed in the literature review. Conversely, another line of reasoning discussed is that there is a high level of strategic dependency between the two types of brands (manufacturer brands and private label), and therefore private label share and related merchandising measures should not go beyond a certain level if delivery of the respective contributions to the categories on the part of manufacturer brands is not to be jeopardised. Thus, there is an existence of equilibrium points in the coexistence of private label and manufacturer brands, beyond which the retailers are reluctant to go with their private label. This latter line of reasoning could be more appealing than the former, because although a shift in the balance of power to retailers may mean that the retailers largely have the final say, if the actual final say is determined by rational business judgement, then the retailers would not likely want to pursue actions that would be detrimental to the product categories. For this reason, given the retail consolidation/concentration and retailer power, as well as the purported link with private label share growth and possibly overdominance, this book investigates the following specific research issues in depth:

Research Issue 1: What is the general nature of the state of balance between private label and manufacturer brands in FMCG product categories?

Research Issue 2a: In a grocery retail landscape characterised by high retail concentration, what is the nature of the state of balance between private label and manufacturer brands in FMCG product categories?

The market share of private label also varies from category to category (ACNielsen 2005; Hoch and Banerji 1993; Nielsen 2014; Sirimanne 2016), and FMCG/supermarket product categories differ on a variety of characteristics such as size, range of competing brands and products (ACNielsen 2005), rate of innovation (Coriolis Research 2002), level of technology (Lehmann and Winer 2002), category commoditisation and related factors. These differences imply that the categories offer different opportunities and challenges, and the retailers would naturally take note of that. Therefore, in addition to the inherent nature of certain categories being prone to relatively higher private label penetration (e.g. commoditised categories), retailer strategic objectives and policies likely differ across the categories. Such objectives and policies may also not be the same across different grocery retailers. Due to the perceived inherent differences in the categories and retail chains, this book investigates the following questions:

Research Issue 2b: How does the balance between private label and manufacturer brands in FMCG product categories compare by category?

Research Issue 2c: How does the balance between private label and manufacturer brands in FMCG product categories compare between grocery retailers?

Product Innovation, Category Marketing Support and Consumer Choice

Both private label and manufacturer brands engage in activities related to product innovation and category support. Experts, however, hold different views on the state of innovation by private label and manufacturer brands, and the state of contribution to category development by the two types of brands (Conn 2005). One perspective is that private label brands are doing a lot on their own in the area of innovation. Performance on innovation is largely influenced by capacity for innovation, and capacity for innovation can be measured in terms of resources, expertise and related output. Retailers have been seen to be boosting their innovative capacity (Lindsay 2004). It has also been reported that private label brands in certain countries have moved away from copying competition to setting their own trends (Silverman 2004). This perspective therefore suggests that private label brands have become masters of their own destinies on innovation. The contrasting view, on the other hand, is that manufacturer brands are leading the way on innovation, since historically, retailers have largely been followers of manufacturer brands on innovation (Aribarg et al. 2014; Coelho do Vale and Verga-Matos 2015; Hoch and Banerji 1993; Olbrich et al. 2016). In addition, it is noted that private label development is not backed by enough research and development money and cannot afford the necessary resources (Conn 2005; Steiner 2004). It is also common knowledge that a supermarket category would normally consist of many different manufacturers who are indeed specialised. Collectively, manufacturer brands are expected to have more resources, skills and knowledge in the areas of product innovation and category support than the retailers.

In addition, related to these two perspectives is the disagreement in the literature on the extent to which private label can develop the categories. One view, consistent with research carried out by Putsis and Dhar (1996), is that private label is capable of expanding category expenditure, developing the market, and not just stealing share from manufacturer brands. A contrary view, however (Anonymous 2005), is that it is the manufacturer brand that actually does expand the categories.

With regard to the two perspectives on manufacturer brand and private label contributions to product innovation and category support, perhaps this depends on the research environment, since private label development is at different stages in different parts of the world and is economy and industry specific. In industries that still have to develop their private label portfolios to the full, manufacturer brands are expected to be doing much more in the areas of innovation and category support than private label brands. The New Zealand grocery retail environment is one such industry. Therefore, it is important to establish, in FMCG/

supermarket industries such as New Zealand, the comparative capacity to innovate and to develop the product categories:

Research Issue 3a: How do private label and manufacturer brands compare on capacity to innovate in the FMCG product categories?

Research Issue 3b: How do private label and manufacturer brands compare on capacity to contribute to category marketing support and development?

The category management exercise is such that retailers are actively involved in managing the categories. Therefore, in the same way as they would be aware of product and category trends relating to sales and profit performance, it is expected that the retailers would be very much aware of the state of affairs regarding manufacturer brand and private label capacity for innovation and support in the product categories.

In addition, researchers are generally in agreement on the importance of innovation as a driving force for the growth of companies and categories (e.g. Anonymous 2004; Brenner 1994; BCG 2005; Doyle and Bridgewater 1998; Guinet and Pilat 1999; Hardaker 1998; Kung and Schmid 2015; Robert 1995). Research has also demonstrated that more innovative categories tend to achieve higher proportions of success than less innovative ones (Booz et al. 1982). Differentiation and brand building through a variety of marketing activities that include advertising are seen as essential ingredients to continued category development.

Furthermore, with respect to grocery product categories, the consumer packaged goods literature has largely portrayed manufacturer brand innovation in relation to private label as a competitive tool that is employed against private label, in addition to competing with other manufacturer brands (e.g. Coriolis Research 2002; Information Resources Inc. 2005; Kumar and Steenkamp 2007b; Verhoef et al. 2002). The literature largely suggests that manufacturer brand innovation does have a negative impact on private label in FMCG categories. The alternative view, of manufacturer brand innovation having a positive impact on private label, has not been investigated in depth. For instance, the imitative and parasitic behaviour of private label (e.g. Collins-Dodd and

Zaichkowsky 1999; Harvey 2000; Ogbonna and Wilkinson 1998) has been reported in the literature, and one would therefore assume that retailers positively benefit from manufacturer brand innovation in this and other related ways. It is therefore reasonable to assume that if a positive impact of manufacturer brand innovation on private label does exist, it would most likely be factored by the powerful retail chains into the determination of policies and strategies on the coexistence of their own brands (private label) with manufacturer brands in the categories. It could be further reasoned that manufacturer brands that are more innovative and category supportive (marketing-wise) are viewed more positively by grocery retailers than those that are not, because the grocery retailers benefit more from them. This discussion therefore warrants an investigation into the following:

Research Issue 4a: What is the state of awareness of FMCG retailers on the comparative capacity of private label and manufacturer brands to innovate and give marketing support to the product categories?

Research Issue 4b: What is the strategic stance of the FMCG retailers with respect to this comparative capacity to innovate and give marketing support to the product categories?

Research Issue 4c: How is this strategic stance related to policies on the coexistence of private label and manufacturer brands in FMCG product categories?

The consumer is the ultimate customer for both types of brands under consideration, that is, private label and manufacturer brands. Issues related to consumer choice are considered in this book to be relevant to the coexistence of manufacturer brands and private label in FMCG/ supermarket product categories as per the reviewed definition of category management, in which consumer focus is a key element. The consumer drives what goes on in the categories (ACNielsen et al. 2006), alongside other strategic factors. Consumers have the freedom to choose (Kaswengi and Diallo 2015; Nelson 2002; Olbrich et al. 2016) and they want brand/ product selection as well. In the category management set-up, the private label brand is seen as being protected by the retailer (e.g. Major and McTaggart 2005), thus retailers may have their own strategic objectives and can employ specific strategic management regimes as they see fit. Yet in the changing grocery retail landscape, where retailers are becoming increasingly powerful due in part to retail consolidation and concentration, it is reasonable to assume that consumer choice issues are still relevant in the determination of the nature of coexistence between manufacturer brands and private label in the categories, even though it is possible that other strategic interests on the part of the retailers would interfere with consumer choice. In connection with this, the following needs to be investigated:

Research Issue 5: What role is played by consumer choice in shaping the coexistence of private label and manufacturer brands in FMCG product categories?

Category Strategic Policies on the Coexistence of Manufacturer Brands and Private Label in FMCG/Supermarket Product Categories

The mode of coexistence between manufacturer brands and private label in the categories is seen in this book as being driven by commonly known factors such as consumer choice. However, other factors that have not been well articulated in the mainstream academic literature and that involve strategic dependency between the two types of brands are considered to be very much at play as well. Due to the fact that it is widely recognised in the literature that power in the FMCG sector has shifted from manufacturers to retailers and that the balance of power is in the hands of the retailers (ACNielsen et al. 2006; Berthon et al. 1997; Hogarth-Scott 1999; Hollingsworth 2004; Hovhannisyan and Bozic 2016; Kumar and Steenkamp 2007a, b; Panigyrakis and Veloutsou 2000; Stanković and Končar 2014; Sutton-Brady et al. 2015; Weitz and Wang 2004), retailers are expected to have a bigger say in how manufacturer brands and private label coexist in the grocery retail categories than manufacturers. It is the author's position that because the balance of power is in favour of the retailers, it is retailer strategic thinking that largely defines manufacturer brand and private label coexistence issues.

As discussed in earlier chapters, such coexistence issues may include: strategic objectives regarding the level to which the private label brand should grow in the categories in terms of share (in relation to the manufacturer brand); strategic decisions on merchandising measures between the two types of brands; competitive strategy aspects of private label in relation to manufacturer brands; how areas of strategic dependency between the two types of brands are handled and how much value is attached to such strategic dependency; and whether recognition for such strategic dependency should be factored into the determination of strategic management regimes governing the coexistence of the two types of brands in the categories. It has also been noted that different categories offer different opportunities and challenges with respect to private label brands. At the same time, some prominent researchers have observed that "the fact that private labels have low share in a category does not imply that a particular retailer cannot create a successful program in that category" (Hoch and Banerji 1993: 66). The retailer would most likely have category-specific policies in place and would be able to influence the nature of manufacturer brand/private label coexistence in the categories instead of just being dictated to by market forces. Therefore, with respect to the mode of coexistence between manufacturer brands and private label, the following needs to be investigated:

Research Issue 6: What is the nature of the coexistence relationship between private label and manufacturer brands in FMCG product categories and how is it driven?

Today's radically altered FMCG landscape (Kumar and Steenkamp 2007a) is characterised by increased retailer power and direct competition between brands owned and managed by owners of the grocery retail shelves and those owned and managed by manufacturers. In this environment, it can be argued that despite the expectations and actions of consumers and manufacturers, retailers have the capacity to significantly influence the final composition of manufacturer brands and private label offered in the supermarket product categories. This is especially so in environments characterised by very high retail consolidation and concentration, where the retailers may be expected to have the capacity to employ coercive power to achieve certain outcomes if they wish.

Dapiran and Hogarth-Scott (2003) have suggested that where there is high retail concentration and low grocery retailer dependence on the supplier, retailers are more likely to employ coercive power. It has been established in the review of the category management literature that in the category management relationship between manufacturers and retailers, each party brings something that is valued by the other party to the table. Therefore, despite the fact that it is widely recognised in the literature that power in the FMCG sector has largely shifted from manufacturers to retailers and that the balance of power is in the hands of the retailers (ACNielsen et al. 2006; Berthon et al. 1997; Hogarth-Scott 1999; Hollingsworth 2004; Hovhannisyan and Bozic 2016; Stanković and Končar 2014; Sutton-Brady et al. 2015; Kumar and Steenkamp 2007a, b; Panigyrakis and Veloutsou 2000; Weitz and Wang 2004), there is still an intricate power–dependence relationship that plays a role in shaping the nature of coexistence between manufacturer brands and private label in FMCG/supermarket product categories. Power and countervailing power are very much at play. Such intricate power relationships can be better understood through an intensive examination of the different sources of power as they relate to the coexistence of the two types of brands in the categories.

The five bases of power (French and Raven 1959; Hunt 2015) can be used to analyse the relationships as there is bound to be an interplay of different sources of power. However, the interplay as it relates to manufacturer brands and private label in the categories is not clear and can therefore be better understood through primary research. Such an investigation would be expected to give insights into the dominant bases of power. On the face of it, coercive power looks to be the more dominant source, but there are issues of strategic dependency between the two types of brands that make the entire power relationship complex. Therefore, the mode of coexistence between manufacturer brands and private label can be analysed from a power perspective and the nature of the power relationship can be better understood through intensive investigation. With this in mind, this book will investigate the following:

Research Issue 7**:** What role is played by power in the mode of coexistence between private label and manufacturer brands in FMCG product categories?

Conceptual Framework

A graphical conceptual framework has been devised for the investigation. A conceptual framework explains the main variables to be studied and the presumed relationships between those variables, and this can be presented in narrative form or graphically (Miles and Huberman 1994), or the two can be made to complement each other, as in this case. Figure 5.1 shows the graphical conceptual framework. The empirical domain that the book explores (Miles and Huberman 1994) is reflected in the combination of the research issues formulated and the graphical conceptual framework devised.

Figure 5.1 can be briefly explained as follows. The environment is the FMCG/supermarket landscape characterised by high retail consolidation and concentration, and direct competition between manufacturer brands and private label, as well as power relationships between category participants (as represented by the lower middle bin in the graphical framework). In this environment, the balance between manufacturer brands and private label in FMCG/supermarket product categories (as represented by the bin on the right) can be influenced by category strategic management regimes (upper middle bin) in a number of areas that include policies on private label growth and share, driving category growth, category management arrangements, shelf matters and retailer brand quality spectrum. Consumer choice (lower left bin), which is a commonly known variable, is factored into the determination of the strategic management regimes governing the coexistence of the two types of brands in the categories. However, at the same time, product innovation and category marketing support are key variables that are also factored into the determination of strategic management regimes governing the coexistence of the two types

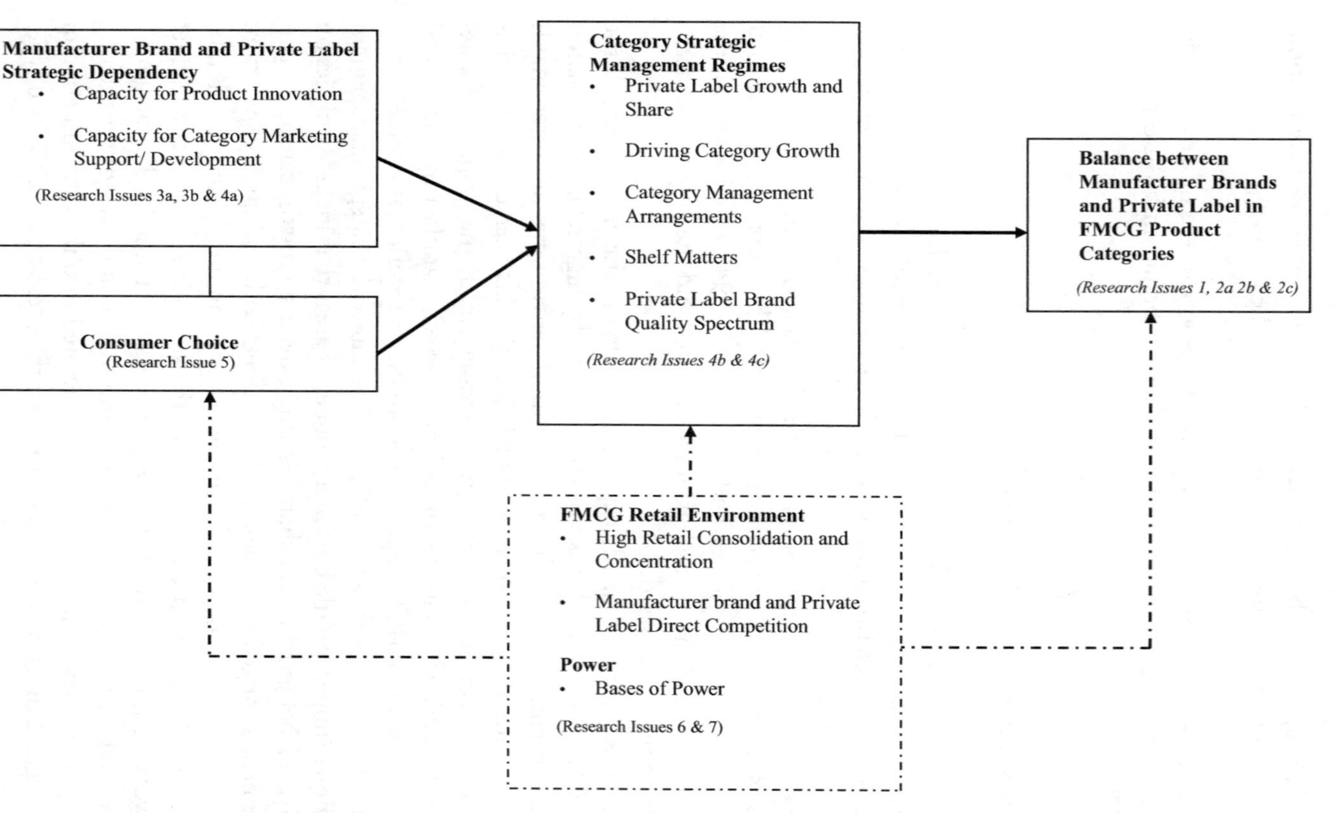

Fig. 5.1 Conceptual framework for the study of private label and manufacturer brands in FMCG product categories

of brands in the categories. Mainstream academic literature has not spelt out the importance of product innovation and category support in this area. In addition, the role played by the bases of power (lower middle bin) is assessed, particularly with regard to which bases are dominant in such an environment in the determination of how the two types of brands should coexist. Given the power imbalances discussed in the literature, one view would be that coercive power is dominant. A converse view that cautiously takes into account the countervailing power of manufacturer brands, on the other hand, would reason that other bases of power are dominant. The true picture will be established through primary research.

Chapter Recap

This chapter has discussed the key aspects of the literature and developed research issues that form the focus of this book. The research issues radiate from the primary research question, which has given rise to three subsidiary research questions. The subsidiary research questions are further distilled into a number of specific research issues under three broad themes that are largely incremental in nature. The research issues are seen as ranging from the "what" and "how" aspects that describe theory to the "why" aspects that explain theory, as is recommended for theoretical contributions (e.g. Whetten 1989). An investigation into these specific research issues will serve to advance the literature in the area of manufacturer brands and private label by demonstrating how these two types of brands coexist in an FMCG/supermarket environment characterised by high retail consolidation and concentration. The next chapter discusses the research paradigm, method and design for this book.

References

ACNielsen. (2005). *The power of private label: A review of growth trends around the world*. New York, NY: ACNielsen.

ACNielsen, Karolefski, J., & Heller, A. (2006). *Consumer-centric category management: How to increase profits by managing categories based on consumer needs*. Hoboken, NJ: John Wiley & Sons Inc.

Anonymous. (2004). Effective innovation. *Strategic Direction, 20*(7), 33–35.
Anonymous. (2005). House brand strategy doesn't quite check out. *The Age*. Retrieved December 3, 2017, from http://www.theage.com.au/news/Business/House-brand-strategy-doesnt-quite-check-out/2005/04/01/1112302232004.html
Aribarg, A., Arora, N., Henderson, T., & Kim, Y. (2014). Private label imitation of a national brand: Implications for consumer choice and law. *Journal of Marketing Research, 51*(6), 657–675.
BCG (Boston Consulting Group). (2005). *Innovation 2005*. Boston, MA: The Boston Consulting Group Inc.
Berthon, P., Hulbert, J. M., & Pitt, L. F. (1997). Brands, brand managers and the management of brands: Where to next? *MSI Report No. 97-122*. Cambridge, MA: Marketing Science Institute.
Booz, Allen & Hamilton. (1982). *New product management for the 1980s*. New York, NY: New York Press.
Brenner, M. S. (1994). Tracking new products: A practitioner's guide. *Research Technology Management, 37*(6), 36–40.
Burt, S. (2000). The strategic role of retail brands in British grocery retailing. *European Journal of Marketing, 34*(8), 875–890.
Burt, S. L., & Sparks, L. (2003). Power and competition in the UK retail grocery market. *British Journal of Management, 14*(3), 237–254.
Chimhundu, R., Hamlin, R. P., & McNeill, L. (2011). Retailer brand share statistics in four developed economies from 1992 to 2005: Some observations and implications. *British Food Journal, 113*(3), 391–403.
Coelho do Vale, R., & Verga-Matos, P. (2015). The impact of copycat packaging strategies on the adoption of private labels. *Journal of Product & Brand Management, 24*(6), 646–659.
Collins-Dodd, C., & Zaichkowsky, J. L. (1999). National brand responses to brand imitation: Retailers versus other manufacturers. *Journal of Product and Brand Management, 8*(2), 96–105.
Conn, C. (2005). Innovation in private label branding. *Design Management Review, 16*(2), 55–62.
Coriolis Research. (2002). *Responding to private label in New Zealand*. Auckland: Coriolis Research.
Cotterill, R. W. (1997). The food distribution system of the future: Convergence towards the US or UK model? *Agribusiness, 3*(2), 123–135.
Dapiran, G. P., & Hogarth-Scott, S. (2003). Are co-operation and trust being confused with power? An analysis of food retailing in Australia and New Zealand. *International Journal of Retail and Distribution Management, 31*(5), 256–267.

Defra (Department for Environment, Food and Rural Affairs). (2006). *Economic note on UK grocery retailing*. London, UK: Department for Environment, Food and Rural Affairs, Food and Drinks Economics Branch.

Doyle, P., & Bridgewater, S. (1998). *Innovation in marketing*. Oxford: Butterworth-Heinemann.

French, J. R. P., & Raven, B. (1959). The bases of social power. In D. Cartwright (Ed.), *Studies in social power*. Institute of Social Research (pp. 150–167), Ann Arbor, MI: The University of Michigan.

Galbraith, J. K. (1952). *American capitalism: The concept of countervailing power*. Boston, MA: Houghton Mifflin Company.

Guinet, J., & Pilat, D. (1999). Promoting innovation: Does it matter? *The OECD Observer*, No. 217/218, Paris, pp. 63–65.

Hardaker, G. (1998). An integrated approach towards product innovation in international manufacturing organizations. *European Journal of Innovation Management, 1*(2), 67–73.

Harvey, M. (2000). Innovation and competition in UK supermarkets. *Supply Chain Management: An International Journal, 5*(1), 15–21.

Hoch, S. J., & Banerji, S. (1993). When do private labels succeed? *Sloan Management Review, 34*(4), 57–67.

Hogarth-Scott, S. (1999). Retailer-supplier partnerships: Hostages to fortune or the way forward for the millennium? *British Food Journal, 101*(9), 668–682.

Hollingsworth, A. (2004). Increasing retail concentration: Evidence from the UK food sector. *British Food Journal, 106*(8), 629–638.

Hovhannisyan, V., & Bozic, M. (2016). The effects of retail concentration on retail dairy product prices in the United States. *Journal of Dairy Science, 99*(6), 4928–4938.

Hunt, S. D. (2015). The bases of power approach to channel relationships: Has marketing's scholarship been misguided? *Journal of Marketing Management, 31*(7–8), 747–764.

Information Resources, Inc. (2005, November). Private label: The battle for value-oriented shoppers intensifies. *Times and trends: A snapshot of trends shaping the CPG industry* (pp. 1–23).

Kaswengi, J., & Diallo, M. F. (2015). Consumer choice of store brands across store formats: A panel data analysis under crisis periods. *Journal of Retailing and Consumer Services, 23*, 70–76.

Kumar, N., & Steenkamp, J. E. M. (2007a). *Private label strategy: How to meet the store brand challenge*. Boston, MA: Harvard Business School Press.

Kumar, N., & Steenkamp, J. E. M. (2007b). Brand versus brand. *International Commerce Review, 7*(1), 47–53.

Kung, H., & Schmid, L. (2015). Innovation, growth, and asset prices. *The Journal of Finance, 70*(3), 1001–1037.

Lehmann, D. R., & Winer, R. S. (2002). *Product management* (3rd ed.). New York, NY: McGraw-Hill.

Lindsay, M. (2004). Editorial: Achieving profitable growth through more effective new product launches. *Journal of Brand Management, 12*(1), 4–10.

Major, M., & McTaggart, J. (2005, November 15). Blueprints for change. *Progressive Grocer*, pp. 89–94.

Miles, M. B., & Huberman, A. M. (1994). *Qualitative data analysis: An expanded sourcebook*. Thousand Oaks, CA: Sage Publications.

Nelson, W. (2002). All power to the consumer? Complexity and choice in consumers' lives. *Journal of Consumer Behaviour, 2*(2), 185–195.

Nielsen. (2014). *The state of private label around the world*. The Nielsen Company.

Ogbonna, E., & Wilkinson, B. (1998). Power relations in the UK grocery supply chain: Developments in the 1990s. *Journal of Retailing and Consumer Services, 5*(2), 77–86.

Olbrich, R., Hundt, M., & Jansen, H. C. (2016). Proliferation of private labels in food retailing: A literature overview. *International Journal of Marketing Studies, 8*(8), 63–76.

Panigyrakis, G., & Veloutsou, C. A. (2000). Problems and future of the brand management structure in the fast moving consumer goods industry: The viewpoint of brand managers in Greece. *Journal of Marketing Management, 16*(1/3), 165–184.

Porter, M. E. (1976). *Interbrand choice, strategy and market power*. Harvard, MA: American Marketing Association.

Putsis, W. P., & Dhar, R. (1996). Category expenditure and promotion: Can private labels expand the pie? *Working Paper*. Yale University, New Haven, CT.

Rizkallah, E. G., & Miller, H. (2015). National versus private-label brands: Dynamics, conceptual framework, and empirical perspective. *Journal of Business & Economics Research, 13*(2), 123–136.

Robert, M. (1995). *Product innovation strategy: Pure & simple*. New York, NY: McGraw-Hill.

Silverman, G. (2004, November 16). Retailers pack new punch in battle with the brands. *Financial Times*.

Sirimanne, E. (2016). Private label: Global, New Zealand & Australian perspectives. *Euromonitor International*. Retrieved September 16, 2017, from www.blog.euromonitor.com

Stanković, L., & Končar, J. (2014). Effects of development and increasing power of retail chains on the position of consumers in marketing channels. *Ekonomika Preduzeća, 62*(5-6), 305–314.

Steiner, R. L. (2004). The nature and benefits of national brand/private label competition. *Review of Industrial Organization, 24*(2), 105–127.

Sutton-Brady, C., Kamvounias, P., & Taylor, T. (2015). A model of supplier–retailer power asymmetry in the Australian retail industry. *Industrial Marketing Management, 51*, 122–130.

Verhoef, P. C., Nijssen, E. J., & Sloot, L. M. (2002). Strategic reactions of national brand manufacturers towards private labels: An empirical study in The Netherlands. *European Journal of Marketing, 36*(11/12), 1309–1326.

Weitz, B., & Wang, Q. (2004). Vertical relationships in distribution channels: A marketing perspective. *Antitrust Bulletin, 49*(4), 859–876.

Whetten, D. A. (1989). What constitutes a theoretical contribution? *The Academy of Management Review, 14*(4), 490–495.

6

Research Paradigm, Research Method and Research Design

Overview

Research issues form the focal theory (Phillips and Pugh 2000: 60) of this book and the issues are based on a review of the literature that constitutes its background theory (p. 59). This chapter discusses the process that was followed in addressing these research issues, and explores data theory (Phillips and Pugh 2000: 61). Specifically, the chapter covers the research paradigm, justification of the case research methodology and research design.

Research Paradigm

This section discusses and justifies the appropriate paradigm for this book, and in the process, alternative paradigms are covered in order to demonstrate the more favourable position of the chosen paradigm in relation to relevant alternatives. A paradigm is a set of connected assumptions about the world that is shared by a community of scientific researchers in their investigation of the world (Kuhn 1962). It can be referred to

R. Chimhundu, *Marketing Food Brands*,
https://doi.org/10.1007/978-3-319-75832-9_6

as a "basic belief system or world-view" that acts as a guide to the researcher (Guba and Lincoln 1994: 105). Four specific objectives that should be accomplished by a paradigm could be outlined as follows. Firstly, guiding professionals in a discipline by way of indicating the important issues and problems confronting the discipline; secondly, developing an explanatory scheme consisting of models and theories, which places the problems and issues in a framework that facilitates solving them; thirdly, establishing appropriate tools consisting of "methodologies, instruments, and types and forms of data collection" (Filstead 1979: 34) to be used in solving puzzles of the discipline; and fourthly, providing an epistemology in which the three tasks covered in the objectives above "can be viewed as organising principles for carrying out the 'normal work' of the discipline" (Filstead 1979: 34). Overall, a paradigm therefore serves as an interpretive framework (Phillips and Pugh 2000) for understanding and explaining the world phenomenon under investigation. In order to justify the appropriate interpretive framework for this book, it is important to look at the elements of a paradigm and how these elements relate to each of the paradigms.

A scientific paradigm has three elements: ontology, epistemology and methodology (Perry et al. 1999). Ontology is reality and its constituent elements (Mitroff and Mason 1982; Perry et al. 1999; Silverman 2004). Epistemology is the relationship between the researcher and reality (Perry et al. 1999), inclusive of the nature and status of knowledge (Silverman 2004) and the philosophy of knowledge, or how people come to know what they know (Deshpande 1983; Mitroff and Mason 1982; Trochim 2000). Methodology is to do with the techniques used by the investigator to discover the reality (Perry et al. 1999). It refers to how the researcher answers the research questions, and includes in addition to the data-gathering techniques, "research design, setting, subjects, analysis, reporting" (Hudson and Ozanne 1988: 508). Although the techniques used by the investigator to discover the reality also deal with how we come to know something, they actually involve the practice of how we come to know rather than the philosophy, which is what epistemology is all about (Trochim 2000). There is therefore a distinction between different paradigms regarding ontology, epistemology and methodology.

Two predominant paradigms in social science research are known as the positivist and interpretive paradigms (Hudson and Ozanne 1988). Each has a range of alternative terms that take on slightly different shades of meaning depending on their use. For instance, the positivist paradigm is also referred to as being scientific, empiricist, quantitative, experimental, deductive (Ticehurst and Veal 2000) or objective (Hudson and Ozanne 1988); and the interpretive paradigm is also referred to as being phenomenological, critical interpretive, qualitative, hermeneutic, reflective, inductive, ethnographic, "action research" (Ticehurst and Veal 2000), naturalistic (Lincoln and Guba 1985), subjective (Rubinstein 1981) or humanistic (ACR Special Session 1985). The interpretive paradigm has three variants: critical theory, constructivism and realism (Guba and Lincoln 1994; Perry et al. 1999). There are therefore four scientific paradigms (inclusive of the interpretive paradigm variants), namely positivism, critical theory, constructivism and realism. These four paradigms, with their philosophical assumptions, form the basic belief systems of alternative enquiry (Perry et al. 1999; Simpson 2003). They are each considered in the context of this research and summarised below.

Positivist Paradigm

Positivism views the world as being external and objective to the investigator; a similar position to that adopted by natural scientists. Researchers are considered to be independent of the research they are undertaking (Ticehurst and Veal 2000). The assumption is that researchers measure independent facts about a single reality that is apprehensible and composed of separate elements whose nature can be known and put into categories (Guba and Lincoln 1994; Tsoukas 1989). The data and their analysis are value free and the data do not change because they are being observed. This scenario has been likened to viewing the world through a "one-way mirror" (Guba and Lincoln 1994: 110), which, in the social sciences, may not offer an adequate picture of the phenomena being investigated. Assumptions that are appropriate for the natural sciences are not necessarily appropriate for social science. For this reason, scholars (e.g. Giddens 1974; Robson 1993) have expressed concerns about the suitability of this paradigm for certain research problems outside the natural science setting.

This study investigates the coexistence of manufacturer brands and private label in FMCG/supermarket product categories; a situation that, from one perspective can be seen as competition, and from another, as cooperation. To address issues set in this mould, it is judged that the researcher cannot achieve the necessary understanding and explanation within a "closed system" as positivism would entail (Perry et al. 1999: 17), but rather must take into account the complex nature of reality and the relevant research problem. They must reflect on, form and revise meanings from managerial experiences and perceptions, as advised by Orlikowski and Baroudi (1991), as well as allow "interaction between theory and fact" (Seale 1999: 21). Given therefore that such research tasks cannot be achieved within the confines of the positivist paradigm, this book will not operate from a positivist perspective.

Phenomenological/Interpretive Paradigm

An approach derived from the social sciences (Remenyi and Money 2004), the interpretive paradigm views the world as socially constructed and subjective. It regards own explanations, i.e. own perspectives, of situations or behaviours by the people being studied as important (Ticehurst and Veal 2000), and the interpretations of the researcher also come into it. Researchers are integral to the research process and seek to understand and find meanings in the broad interrelationships of situations being investigated; therefore, aspects related not only to what is happening but also to why it is happening are central to the research task (Saunders et al. 1997; Ticehurst and Veal 2000). The three variants of this paradigm, however, are suited to different research situations.

Critical Theory

Critical theory research seeks to critique and transform social, economic, political, cultural, ethnic and gender values. The research enquiries of critical theorists often involve long-term historical and ethnographic studies of organisational processes and structures (Perry et al. 1999). Marxists, feminists and action researchers are typical examples of critical theorists (Neuman 2000; Perry et al. 1999; Seale 1999). Critical theory assumptions are fundamentally subjective and hence knowledge is grounded in historical and social routines, and is therefore not value free but value dependent. The critical theorist paradigm is not appropriate for research in marketing research since market researchers rarely (if at all) have the aim of being transformative intellectuals who liberate people from their inherent historical mental, emotional and social structures (Guba and Lincoln 1994). The book aims to better understand the coexistence of manufacturer brands and private label in FMCG product categories from the viewpoint of the nature of coexistence, management decisions taken, underlying motives and relevant theories. The research does not intend to "change the world" (Neuman 2000: 76) by way of intellectually transforming the FMCG sector's manufacturer brand/private label business strategy in the manner that Marxists or feminists, for instance, would want to change the political and/or social world. The critical theorist paradigm is therefore not appropriate for this research book.

Constructivism

Constructivists hold the view that reality is a construct in the minds of individuals. This ontological position makes room for an infinite number of constructions, and as a result, multiple realities are deemed to exist (Lincoln and Guba 1985). Meaning is seen as having more value than measurement as perception itself becomes the most important reality. Moreover, like critical theory, constructivism enquires about the relevant ideologies and values behind the findings (Guba and Lincoln 1994; Perry et al. 1999).

The constructivist approach is deemed to be suitable for social science research topics such as beauty and religion (Hunt 1991). However, the approach is largely inappropriate for business research because it disregards concerns about the economic and technological aspects of business (Hunt 1991; Perry et al. 1999). This book takes into account economic, technological and social dimensions related to the balancing of manufacturer brands and private label in FMCG/supermarket product categories. The constructivist paradigm would therefore not serve as a suitable paradigm for the research.

Realism

The realist paradigm, which is at times termed the critical realist or post-positivist paradigm (Guba and Lincoln 1994), differs from, but has some commonalities with, both positivism and constructivism (Perry et al. 1999). Realists hold the ontological position that there is a real world to discover but the discovery cannot be achieved in a perfect or clear-cut manner (Guba and Lincoln 1994; Perry et al. 1999; Tsoukas 1989). Realists do not consider perception per se to be reality as constructivists and critical theorists would do, but instead see perception as "a window on to reality through which a picture of reality can be triangulated with other perceptions" (Perry et al. 1999: 18). Triangulation presents the best hope of achieving objectivity and thereby obtaining a better picture of the reality (Perry et al. 1999; Trochim 2000).

Three domains of reality, namely mechanisms, events and experiences, can be identified in the realist world (Bhasker 1978), and these constitute a combination of observable and non-observable phenomena. The discovery of such observable and non-observable structures and mechanisms underlying events and experiences is the objective of realist research (Tsoukas 1989). The task of the investigator is therefore to discover and identify, then describe and analyse those structures and generative mechanisms related to the phenomena being investigated. A direct cause and effect relation is not the prime objective. The main concern is that of unveiling and exploring the underlying causal tendencies or powers of the phenomena under investigation (Bhasker 1978).

This book investigates research questions related to the coexistence of manufacturer brands and private label in FMCG/supermarket product categories. This calls for the exploration of observable phenomena such as the shelf space and category share situations of the brands, as well as non-observable phenomena such as the underlying causal tendencies, both of which are in fact key characteristics of the realist paradigm. It is also reasonable to assume that, in determining the nature of coexistence between manufacturer brands and private label, managers in the FMCG sector make rational business decisions taking into account economic, technological and other dimensions. The realist paradigm does not exclude such business dimensions. The paradigm also makes room for the incorporation of a variety of sources of evidence in tackling the not-so-obvious relationships between manufacturer brands and private label in FMCG categories in the radically changed consumer packaged goods landscape. The author of this book therefore judges that realism is the most appropriate paradigm for the book because its ontological and epistemological standpoints are more in line with the requirements of the research topic than any of the other paradigms.

Justification of Case Research Methodology

The preceding section argued for the realist variant of the interpretive paradigm as the appropriate philosophical framework for this study. This section discusses the case study research methodology, which is judged to be the appropriate research methodology for the book. A technical definition of case study research that incorporates the methodology's scope and key characteristics is given by Yin (2003: 13–14) as:

> an empirical enquiry that investigates a contemporary phenomenon within its real life context, especially when the boundaries between phenomenon and context are not clearly evident [...] The case study enquiry copes with the technically distinctive situation in which there will be many more variables of interest than data points, and as one result; relies on multiple sources of evidence, with data needing to converge in a triangulating fashion, and as another result; benefits from the prior development of theoretical propositions to guide data collection and analysis.

The case study methodology is seen as a comprehensive, "all-encompassing method" that covers the logic of research design, techniques of data collection and specific data analysis approaches (Yin 2003:

14) rather than just a restricted feature such as the data collection method alone (Stoecker 1991). A number of traditional social science methodologies that include, among other things, experiments, surveys, case studies (Robson 1993; Yin 2003), archival analysis and histories (Yin 2003) have been considered in the process leading to the choice of the case study methodology. All of these methods have been employed successfully in various marketing academic research projects that were best suited to each method: for instance, experimental study (e.g. Hamlin 1997), survey method (e.g. Lindblom and Olkkonen 2006), case study (e.g. Lindgreen 2001), archival analysis (e.g. Hoch et al. 2002a, b) and historical study (e.g. Low and Fullerton 1994).

This book largely focuses on the "whats", "hows" and "whys" of the coexistence of manufacturer brands and private label in FMCG product categories, and the major reasons for not adopting some of the methodologies mentioned above are briefly outlined as follows. The historical method would not be appropriate as the primary methodology because the research topic focuses largely on a contemporary issue (Yin 2003), even though the research still makes use of some historical data on manufacturer brand and private label share trends. Additionally, the researcher could gain access to the managers and obtain their stories on the research issues at hand. Archival analysis would not be appropriate as the sole or main methodology because answers as to why certain things are happening regarding the coexistence of manufacturer brands and private label would be hard to establish. However, some of the historical data on private label trends come from archival records. The survey method would not be appropriate for more or less the same reason. It does not allow for the acquisition of rich data on why (Yin 2003) managers make certain decisions in balancing manufacturer brands and private label. At the same time, an experiment would not be appropriate as it requires control of behavioural events (Yin 2003). There is no scope in a study of this nature to manipulate the behaviour of the retail chain managers and manufacturing company managers concerned.

Turning to the preferred approach for this book, which is the case study methodology, several interrelated factors make this approach comparatively more favourable. The first factor is to do with the nature of research questions (Yin 2003) and the resultant need to "gain a deep understanding" (Perry et al. 1999: 20) of managerial decisions relating to manufacturer brands and private label in FMCG product categories. It is important that the method of research be appropriate for the nature of the research question (Silverman 2004). Research issues developed in Chap. 5 are decomposed from the primary research question. Research issues 1 and 2 largely reflect the "whats" and "hows" of the coexistence of manufacturer brands and private label in the categories. Research issues 3, 4, 5, 6 and 7 largely reflect the "hows" and "whys" of the coexistence, including related management decisions. To a great extent, the research tackles "how" and "why" questions and the case research methodology has the capacity to do justice to such research questions (Robson 1993; Yin 2003). The qualitative case study method allows the researcher to acquire deep and detailed qualitative data by getting closer to the phenomenon physically and psychologically through in-depth interviews (Perry et al. 1999), enabling a better understanding of the researched phenomenon (Gilmore and Carson 1996; Perry et al. 1999). Furthermore, case research can "draw on a wider array of documentary information, in addition to conducting interviews" (Yin 2003: 6) to address the "why" research question.

The second factor supporting the use of case research is that the case method "is preferred in examining contemporary events, but when the relevant behaviours cannot be manipulated" (Yin 2003: 7). The researcher in this case has no control over the behaviours of FMCG participants, but indeed gains an understanding of what is happening, how and why, without changing the contextual environment of the research or influencing managerial decisions and their underlying rationale. In addition, the coexistence of manufacturer brands and private label in FMCG product categories is a contemporary and ongoing phenomenon which is seen to be unfolding from the historical past.

The third justification for employing case research in the investigation of manufacturer brands and private label in FMCG product categories is that, with regard to research that relates to pre-paradigmatic stages of a phenomenon (Borch and Arthur 1995), theory-building via case studies can play an important part (Perry et al. 1999) in expanding the frontiers of knowledge. Some research experts in the case study methodology have categorically asserted that "building theory from case study research is most appropriate in the early stages of research on a topic or to provide freshness in perspective to an already researched topic" (e.g. Eisenhardt 1989: 548). While the coexistence of manufacturer brands and private label in consumer packaged goods categories is not an entirely new phenomenon, the radically altered FMCG landscape characterised by extreme retail consolidation and concentration, increased emphasis on private label brands as part of retail marketing strategy, increased employment of information technology, and increased emphasis on category management, presents a research topic that is transforming significantly and exhibiting pre-paradigmatic characteristics to the extent of seriously requiring an investigation that provides a fresh perspective. The particularly unprecedented nature (by world standards) of retail consolidation and concentration in New Zealand presents fertile ground for the employment of case study research to tackle manufacturer brand/private label related issues. Academic work in the area of private label by Kumar and Steenkamp (2007), considered by its authors to be the "first book to deal with a radically altered landscape" (back flap) further supports the near pre-paradigmatic nature of the research area and the need for new interpretations. The elements of newness in the changed landscape also make the case study investigative approach appropriate for this book.

The fourth aspect supporting use of the case study method is related to the above factors and involves the required classification of researched phenomenon into categories and the identification of relevant interrelationships among those categories (Perry et al. 1999) as a process necessary for successful theory-building. Through case study research, it is possible to isolate categories, define them precisely and then determine the relationships between them (Perry et al. 1999). The whole process is considered to be vital in handling the complex nature of organisational operations and managerial experiences (Bonoma 1985; Gilmore and Carson 1996) related to the balance between manufacturer brands and private label, capacity and incentives for innovation and category development, and strategic management regimes governing the coexistence of manufacturer brands and private label in FMCG/supermarket product categories.

Research Design

Theory-Building Nature of the Research

This book comprises theory-building research seeking to better understand how manufacturer brands and private label coexist in FMCG/supermarket product categories in the highly concentrated consumer packaged goods landscape, and why. Theory-building research, in addition to being suitable for new research areas, also suits research topics in which a certain level of understanding has been achieved already, but where more should be done in terms of theory-building before theory testing can take place (Miles and Huberman 1994; Perry 1998). Although some academic research has been carried out on manufacturer brands and private label, the radical shift in the FMCG landscape introduces new dimensions that support a fresh examination. This theory-building exercise therefore involves a situation where "elements of the theory are being confirmed or disconfirmed, rather than being tested for generalisability to a population" (Perry et al. 1999: 20).

A theory is indeed "a statement of relationships between units observed or approximated in the empirical world" (Bacharach 1989: 498). In building the theory, the researcher "interweaves a story (the theory)" (Neuman 2000: 40) about the coexistence of manufacturer brands and private label in FMCG product categories in a highly concentrated grocery retail landscape with what can be observed when the researcher "examines it systematically (the data)" (Neuman 2000: 40) through qualitative case study research. The "story" in this case is the set of research issues (and conceptual framework) advanced in Chap. 5. The theory will cover building blocks suggested by a number of authorities in theory-building research, who have identified four essential elements of a complete theory. These are summarised in Table 6.1.

The various concepts that constitute the conceptual framework and research issues of this book cater for the "what" aspects of the theory. And, as mentioned in the preceding section, "what" and "how" elements are covered by research issues 1 and 2. The "why" element is catered for in research issues 3 to 7, although the "how" aspect is still interwoven

Table 6.1 Building blocks of theory development

Element(s) of theory	Role of element(s)	Constituent factors of the elements	Broad components of the theory	Criteria for judging appropriateness of theory
What	Identifying and/or describing	Variables/constructs/concepts	Domain/subject	Completeness/ comprehensiveness; parsimony
How	Describing	Relationships among the variables/ constructs/concepts	Domain/subject	Logic; reasonableness
Why	Explaining	Underlying psychological, economic or social dynamics giving rise to the variables/constructs/concepts and the relationships between them	Rationale	Logic; reasonableness
Who; where; when	Setting limits/ bounding	Temporal and contextual factors	Context	Range of theory; boundaries of generalisability

Source: Table created for this book based on Dubin (1978) and Whetten (1989)

into these research issues/questions. The last elements (who, where, when) that set boundaries to the theory are partly covered by reference that has already been made to the highly concentrated FMCG industry, and are further taken care of later in case selection, and in the limitations of the case study methodology. Eisenhardt (1989) has identified the final products of building theory from case study research as being concepts, a conceptual framework, propositions or mid-range theory. The final product of this research book is a theoretical framework and research propositions. In the theory development exercise, the research issues and preliminary conceptual framework will be tested for their adequacy using empirical, case study data from the FMCG/supermarket industry.

This book recognises the importance of prior theory as its research questions and issues have been informed by discussions in the literature chapters (Chaps. 2, 3 and 4) and research issues chapter (Chap. 5). The development of theory before collecting case study data is an essential step in undertaking case study research (Yin 2003). The main reason for making use of prior theory in the area of manufacturer brands and private label in FMCG product categories is to focus the research and avoid being overwhelmed by the volume of unnecessary data (Eisenhardt 1989). Research questions and specific research issues, which are part and parcel of prior theory, direct attention to issues that are studied, so the research avoids the danger of attempting to cover everything. The use of prior theory enables the focus to be on "facets of the empirical domain that the researcher most wants to explore" (Miles and Huberman 1994: 23). The research is both theoretical and empirical (Trochim 2000); theoretical in the way it develops theory about manufacturer brands and private label, and empirical in the way it involves collecting and analysing secondary and primary empirical data about what is happening in the categories.

Research experts have debated induction and deduction as alternative case study research approaches, with some researchers (e.g. Glasser and Strauss 1967) claiming initially that induction (particularly grounded theory) is superior because there is no possibility of the researcher being influenced by a given theory under consideration, but with one of those researchers later taking a middle of the road approach, recognising that a mix of induction and deduction is preferable (Strauss 1987). The view that induction and deduction are in fact linked and complementary

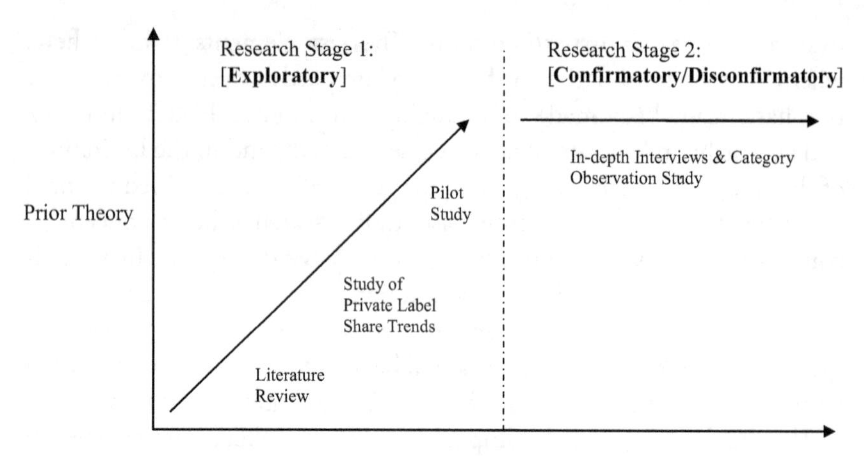

Fig. 6.1 Two-stage theory-building research process. Source: Adapted from Perry (1998)

approaches is shared by a number of researchers (e.g. Miles and Huberman 1994; Parkhe 1993; Perry 1998; Richards 1993; Ticehurst and Veal 2000; Trochim 2000; Yin 2003). In this study, pure induction would prevent the book from benefiting from existing theory, and pure deduction would prevent the development of new and useful theory (Perry 1998). The research therefore involves a continuous interplay between induction and deduction because both extremes are considered to be "untenable and unnecessary" (Parkhe 1993: 253). The research benefits from the interplay between data and theory.

A two-stage theory-building research process was adopted as illustrated in Fig. 6.1. The figure also illustrates the order and process of the research tasks undertaken. The literature review has resulted in the generation of specific research issues, followed by a preliminary study involving the collection and analysis of the private label share data of four developed economies, and then a largely New Zealand based pilot study which was conducted in preparation for the main study.

Quality Issues in Case Study Research

This book recognises the fact that quality is of the utmost importance in research (Trochim 2000), therefore appropriate quality criteria are applied in the conduct of the investigation. While the research made use of some quantitative data (e.g. private label share, shelf space measurements and so on), the research is largely interpretive in nature. With such interpretive research, it has been noted (Summers 2001) that some qualitative researchers seem to feel that there is naturally less obligation on their part to comply with all aspects related to rigorous quality criteria, a view that this research does not subscribe to.

The conventional criteria for establishing quality of research are validity and reliability. These criteria have their roots in quantitative research, but they can be flexed to provide suitable quality criteria for interpretive research as well. Ticehurst and Veal (2000) have defined validity as "the extent to which the data collected truly reflect the phenomenon being studied" (p. 23); reliability as "the extent to which research findings would be the same if the research were to be repeated at a later date, or with a different sample of subjects" (p. 24); and generalisability (which is part of validity) as "the probability that the results of the research findings apply to other subjects, other groups, and other conditions" (p. 24). Within the broad framework of validity and reliability as quality criteria, four quality tests, i.e. construct validity, internal validity, external validity and reliability, are considered to be applicable to all social science research methodologies, including case study research (Yin 2003) such as this one. These quality criteria are applied throughout the case research process and not just at the start of the project. How the quality criteria are addressed in this book is summarised in Table 6.2.

Use of the terms construct validity, internal validity, external validity and reliability has long been associated with the positivist paradigm and is considered by some scholars not to accurately reflect quality issues related to qualitative research. Alternative terms for judging qualitative research have been suggested as credibility (for internal validity), transferability (for external validity), dependability (for reliability) and confirmability (for objectivity) (Guba 1981).

Methodologists have debated the value and legitimacy of these alternative standards for judging qualitative research. Many quantitative researchers perceive the alternative criteria as a mere relabelling of the successful quantitative criteria in order to give greater legitimacy to qualitative research. Quantitative researchers suggest that a correct read-

Table 6.2 Case research tactics for four quality criteria

Quality criteria	Case study tactics employed in this research to address respective quality criteria	Phase of research in which tactics occur
Construct validity	Use of multiple sources of evidence, namely secondary data, in-store category observational data and in-depth interview data. Data triangulation and methodological triangulation employed.	Data collection
	Establishing the following chain of evidence: documented prior theory (with derived specific research issues and conceptual framework); citations of evidentiary sources; case study database and case study report.	Research design, data collection and composition
	Research participant validation (i.e. respondent validation) of research interview data/ transcripts and category observational data.	Composition
Internal validity	Searching for the "hows" and "whys" behind relationships and outcomes (Eisenhardt 1989); explanation building; use of contextual descriptions; making comparisons with the literature; addressing rival explanations.	Research design, data analysis and composition
External validity	Purposive sampling to focus on theoretically useful FMCG categories.	Research design
	Use of replication logic in multiple case studies (five FMCG categories and four supermarket groups under two retail chains researched; both literal and theoretical replication used). Seeking to achieve analytic generalisation (and possibly case-to-case transfer).	Research design
	Comparing evidence with extant literature (to uncover commonalities and areas of conflict, then push for generalisation across cases)	Data analysis
Reliability	Use of case study protocol (include field procedures, observational study form, in-depth interview protocol and structure of report).	Data collection
	Developing a formal case study database consisting of organised, categorised and complete notes, case documents, tabular materials and narratives.	Data collection

Source: Adapted from Yin (2003: 34), Eisenhardt (1989) and Parkhe (1993) for this book

ing of the conventional criteria would show that they are not just limited to quantitative research, but can be applied to qualitative data equally well. They further argue that the alternative criteria reflect a different philosophical perspective that is in fact subjectivist rather than realist in nature (Trochim 2000). Other researchers have therefore reclassified Guba's (1981) terms as a constructivist typology and suggested the following separate terms for the realist paradigm: ontological appropriateness (for construct validity), contingent validity (for internal validity), multiple perceptions of reality (for external validity), methodological trustworthiness (for reliability) and analytical generalisations (Simpson 2003).

This research prefers to lay more emphasis on the tactics that are used to achieve high quality interpretive research. Thus, it advocates the broadening of the conventional criteria (Trochim 2000) to include quality tactics for qualitative research, and notes that Yin (2003) has, in a way, actually done that. In this regard, the research takes the position that most of the case study tactics suggested by Yin (2003) are largely applicable to the interpretive case study research on manufacturer brands and private label in FMCG product categories, and will be complemented by other suggested tactics for the realist paradigm. Additionally, the book makes use of a blanket checklist of the characteristics of high quality case research as outlined in Table 6.3.

Criteria for Case Selection

This section justifies a multiple-case design for the research, and on the basis of this design goes on to explain the choice of specific cases, the number of cases and the number of interviews conducted. While manufacturer brands belong to various manufacturing companies and private label belongs to retail chains, the coexistence of the brands manifests itself in FMCG/supermarket categories (on the shelves and/or in fridges/freezers). Different types of brands and products feature in different categories alongside brands and products of a similar nature. This book takes the category as the case and primary unit of analysis.

Table 6.3 Characteristics of high quality case research and how they are employed in this book

Characteristic	Application to the research on manufacturer brands and private label in FMCG product categories
Case study as a story	Individual case reports are presented for each of the categories and supermarket groups studied before cross-case analysis takes place.
Draws on multiple sources of evidence	Study draws on secondary data, in-store category observational data and in-depth interview data.
Triangulation of sources of evidence	Data and methodological triangulation employed.
Should provide meaning in a context	Contextual descriptions of the categories and retailers provided.
Shows both in-depth understanding of central issue(s) being explored and broad understanding of related issues and context	Prior theory used in the research. Key issues derived from prior theory and experiences are outlined in the preliminary conceptual framework and specific research issues. Contextual environment of the categories and supermarket groups integrated.
Has a clear-cut focus on either an organisation, a situation or a context	Clear focus on manufacturer brands and private label in specific FMCG categories.
Should be reasonably bounded (should not stretch over too wide a canvas, either temporal or spatial)	Bounding covers, inter alia, New Zealand FMCG/ supermarket industry, manufacturer brands and private label, and selected categories. Further bounding criteria spelt out in case selection.
Should not require the researcher to become too immersed in the object of the research	The closest the researcher came to the object of the research was during in-depth interviews and in-store category observation studies. Participant observation was not employed.
May draw on either qualitative or quantitative tools, or both, for evidence collection and/or analysis but will not be exclusively quantitative	Includes quantitative data on private label brand share and shelf data for the respective categories, but the bulk of the research is qualitative in nature.
Needs to have a thoroughly articulated protocol	Interview protocol based on research issues developed, as well as additional insights from secondary data and pilot study.

Source: Table created for this book based on Remenyi and Money (2004: 73–74)

Rationale for Multiple-Case Design

The rationale for employing multiple-case design (i.e. studying more than one category) rather than single-case design (i.e. studying only one category) is determined in this research on the basis of the capability of the design to adequately address the research issues. Case study researchers have taken different positions on the issue of single versus multiple-case designs. The single-case design is supported by Dyer and Wilkins (1991) as being capable of generating rich theoretical insights, since it enables a case to be studied in depth and focuses on telling the story of that particular case. Generalising in a single-case study is on the basis of a match to underlying theory (Miles and Huberman 1994), which may have been developed from prior theory. Other researchers have, however, supported and encouraged the use of multiple-case design, where the nature of the research topic allows for such design (Yin 2003). Conducting the classic case study (single-case) is seen as conducting a single experiment. Multiple cases should be seen as "multiple experiments" based on "replication logic" (Yin 2003: 47). Evidence based on multiple cases is considered to be more compelling and therefore makes the study more robust (Eisenhardt 1989; Herriott and Firestone 1983; Miles and Huberman 1994; Yin 2003). In addition, according to Perry (1998: 792), "several cases should usually be used in postgraduate research because they allow cross-case analysis to be used for richer theory building".

To make the evidence from the study on manufacturer brands and private label more compelling and to allow cross-case analysis for the building of richer theory, a multiple-case design is adopted. This means that while the research still generalises to each case on the basis of a match to underlying theory, the study is made more robust by the additional dimension of multiple-case design that enables "generalising from one case to the next on the basis of a match to underlying theory" (Miles and Huberman 1994: 29), thereby adding "confidence to findings" (Miles and Huberman 1994: 29). Both literal replication (predicting similar results) and theoretical replication (predicting contrasting results for predictable reasons) (Yin 2003) are achieved. The chosen cases (categories) and embedded cases within them have "multi-dimensional blends of theoretical and literal replication" (Perry 1998: 794). It is noted that some researchers have used up to three dimensions of theoretical replication (Perry 1998).

Rationale Behind the Choice of Cases

Cases were selected from the ACNielsen (2005) list of product categories. The following categories were chosen for the research: milk, flour, cheese, breakfast cereals and tomato sauce. The purposive sampling technique was used to select cases (i.e. categories), allowing the researcher to choose cases because they illustrate features or processes that the research is concerned with (Silverman 2004). This type of sampling suits a qualitative case study research of this nature because it enables the researcher to seek out categories where "the processes being studied are most likely to occur" (Denzin and Lincoln 1994: 202). The choice of cases is therefore theory-driven (Miles and Huberman 1994), ensuring the selection of information-rich cases (Patton 1990; Perry 1998; Saunders et al. 1997). Selecting categories in such a purposeful manner makes the research "an information-rich case study in which to explore" specified research issues (Saunders et al. 1997: 142). The range of categories chosen includes a mix of categories that have similarities and differences in the following characteristics: size (volume/value); private label penetration; category development activities; product, packaging and branding related innovation; role of technology in the category; category commoditisation; and trust of the brand. This has enabled the research to yield meaningful insights through replication logic. Each of the specific research issues outlined in Chap. 5 was systematically investigated by way of testing for adequacy. Taking the selected categories into account, the design adopted is given in Table 6.4. The design has three dimensions of theoretical replication.

Rationale for Number of Cases and Number of Interviews

This book takes the stance that the exact number of cases studied should depend on the specific requirements of the research being conducted. For classic case designs, there is no question about what the number of cases should be as the number is always a single case, although it can have embedded cases within. However, for research that has scope for multiple-case design, a decision has to be based on the actual number of cases. In the literature, there are varying views on the issue, ranging from views

Table 6.4 Research design employed

Dimensions of literal and theoretical replication	Cases and embedded cases
Dimension 1: Case(s) (primary unit of analysis)	Category (five categories) - Milk - Flour - Cheese - Breakfast cereals - Tomato sauce
Dimension 2: Embedded cases (embedded unit of analysis 1)	Retail chain (the two retail chains)
Dimension 3: Embedded cases (embedded units of analysis 2 and 3)	Supermarkets (two supermarket chain stores under each retail chain) Manufacturers (whose brands feature in the respective categories; at least two chosen for each category)

Source: Table created for this book

that only give criteria without suggesting a specific number (e.g. Lincoln and Guba 1985; Patton 1990; Romano 1989) to those that make suggestions on the approximate number of cases (Eisenhardt 1989; Hedges 1985; Miles and Huberman 1994; Yin 2003).

It has been noted, based on the latter group of authors' suggestions, that the recommended range falls between two to four cases as the minimum and ten to fifteen cases as the maximum (Perry 1998). Furthermore, according to Perry et al. (1999: 19), "the rigorously analytical method of case study research [is] usually based on many interviews within 4 to 14 cases" undertaken using an interview protocol. With the generally recommended range in mind, and the observation made that "with fewer than four cases, it is often difficult to generate theory with much complexity, and its empirical ground is likely to be unconvincing" (Eisenhardt 1989: 545), this research makes use of the five cases or categories mentioned—that is, milk, flour, cheese, breakfast cereals and tomato sauce.

As can be seen from Table 6.4, each of the five categories has embedded cases at retail chain level and at supermarket chain level. Investigating more than five categories would make the research unwieldy, especially taking into account the embedded cases. For each category, in-depth interview data came from retail chain head office management, private label company management, supermarket chain management, manufacturer/supplier company management and industry consultants. As will be noted in the chapter on private label and manufacturer brand research execution, Chap. 7, at retailer level, more interviews were conducted with one retail chain than the other as the second retail chain was included to provide an opportunity for a check and triangulation. In addition, it had to do with limited access.

Given the number of cases and embedded cases, as well as the respective units of data collection, the research made use of 46 interviews (with 49 interview participants: two interviews being held jointly, one with three managers, and the other with two managers). The number of interviews conducted falls within the range of interviews recommended for the highest level of academic research using the case study methodology, by experienced researchers. Based on experience and anecdotal evidence, Perry (1998: 794) has noted that such high-level research "requires about 35 to 50 interviews".

Number of In-Store Category Observation Study Forms Completed

A category observation study was carried out in 18 supermarket stores; these were the stores where in-depth interviews were conducted. The study involved establishing the following for manufacturer brands and private label, through measurement and general observation: shelf space, shelf facings, shelf position, number of brands and number of products. Five forms were completed in each store for each of the five main categories of interest. Again, as will be noted in the research execution chapter, Chap. 7, more category observation exercises were carried out in one retail chain in relation to the other for the same reasons given in the rationale for the number of cases and number of interviews.

Chapter Recap

This chapter has covered the research paradigm, justification of the case methodology and research design.

The next chapter (Chap. 7) looks at private label and manufacturer brand research execution.

References

ACNielsen. (2005). *The power of private label: A review of growth trends around the world*. New York, NY: ACNielsen.

ACR Special Session. (1985). Applying humanistic methods to consumer research: Four practical examples. In R. J. Lutz (Ed.), *Advances in consumer research* (Vol. 13, pp. xiv–xxv). Provo, UT: Association of Consumer Research.

Bacharach, S. B. (1989). Organisational theories: Some criteria for evaluation. *The Academy of Management Review, 14*(4), 496–515.

Bhasker, R. (1978). *Realist theory of science*. Harvester: Wheatsheaf.

Bonoma, T. V. (1985). Case research in marketing: Opportunities, problems and a process. *Journal of Marketing Research, 22*(2), 199–208.

Borch, O. J., & Arthur, M. B. (1995). Strategic networks among small firms: Implications for strategy research methodology. *Journal of Management Studies, 32*(4), 419–441.

Denzin, N., & Lincoln, Y. (Eds.). (1994). *Handbook of qualitative research* (2nd ed.). Thousand Oaks, CA: Sage.

Deshpande, R. (1983). Paradigms lost: On theory and method in research in marketing. *Journal of Marketing, 47*(4), 101–110.

Dubin, R. (1978). *Theory development*. New York, NY: Free Press.

Dyer, W. G., Jr., & Wilkins, A. L. (1991). Better stories, not better constructs, to generate better theory: A rejoinder to Eisenhardt. *The Academy of Management Review, 16*(3), 613–619.

Eisenhardt, K. M. (1989). Building theories from case study research. *The Academy of Management Review, 14*(4), 532–550.

Filstead, W. J. (1979). Qualitative methods: A needed perspective in evaluation research. In T. D. Cook & C. S. Reichardt (Eds.), *Qualitative and quantitative methods in evaluation research* (pp. 33–48). Beverly Hills, CA: Sage Publications.

Giddens, A. (Ed.). (1974). *Positivism and sociology*. London: Heinemann.

Gilmore, A., & Carson, D. (1996). Integrative qualitative methods in a services context. *Marketing Intelligence & Planning, 14*(6), 21–26.

Glasser, B. G., & Strauss, A. L. (1967). *The discovery of grounded theory: Strategies for qualitative research*. Chicago, IL: Aldine Publishing Company.

Guba, E. G. (1981). Criteria for assessing the trustworthiness of naturalistic enquiries. *Educational Communication and Technology Journal, 29*(2), 75–91.

Guba, E. G., & Lincoln, Y. S. (1994). Competing paradigms in qualitative research. In N. K. Denzin, & Y. S. Lincoln, Y. S. (Eds.), *Handbook of qualitative research* (pp. 105–117). Thousand Oaks, CA: Sage Publications.

Hamlin, R. P. (1997). *The meat purchase decision: An experimental study*. PhD thesis, University of Otago, Dunedin, New Zealand.

Hedges, A. (1985). Group interviewing. In R. Walker (Ed.), *Applied qualitative research* (pp. 71–91). Aldershot: Gower Publishing Company.

Herriott, R. E., & Firestone, W. A. (1983). Multisite qualitative policy research: Optimising description and generalizability. *Educational Researcher, 12*(1), 14–19.

Hoch, S. J., Montgomery, A. L., & Park, Y. H. (2002a). Why private labels show long-term market evolution. *Marketing Department Working Paper*, Wharton School, University of Pennsylvania, PA.

Hoch, S. J., Montgomery, A. L., & Park, Y. H. (2002b). Long-term growth trends in private label market shares. *Marketing Department Working Paper #00-010*, Wharton School, University of Pennsylvania, PA.

Hudson, L. A., & Ozanne, J. L. (1988). Alternative ways of seeking knowledge in consumer research. *Journal of Consumer Research, 14*(40), 508–521.

Hunt, S. (1991). *Modern marketing theory*. Cincinnati, OH: South-Western.

Kuhn, T. (1962). *The structure of scientific revolutions*. Chicago, IL: University of Chicago Press.

Kumar, N., & Steenkamp, J. E. M. (2007). *Private label strategy: How to meet the store brand challenge*. Boston, MA: Harvard Business School Press.

Lincoln, Y. S., & Guba, E. G. (1985). *Naturalistic enquiry*. Beverly Hills, CA: Sage Publications.

Lindblom, A., & Olkkonen, R. (2006). Category management tactics: An analysis of manufacturers' control. *International Journal of Retail and Distribution Management, 34*(6), 482–496.

Lindgreen, A. (2001). A framework for studying relationship marketing dyads. *Qualitative Market Research: An International Journal, 4*(2), 75–87.

Low, G. S., & Fullerton, R. A. (1994). Brands, brand management and the brand manager system: A critical-historical evaluation. *Journal of Marketing Research, 31*(2), 173–190.

Miles, M. B., & Huberman, A. M. (1994). *Qualitative data analysis: An expanded sourcebook*. Thousand Oaks, CA: Sage Publications.

Mitroff, I. I., & Mason, R. O. (1982). Business policy and metaphysics: Some philosophical considerations. *Academy of Management Review, 7*(3), 361–371.

Neuman, W. L. (2000). *Social research methods: Qualitative and quantitative approaches* (4th ed.). Boston, MA: Allyn and Bacon.

Orlikowski, W. J., & Baroudi, J. J. (1991). Studying information technology in organisations: Research approaches and assumptions. *Information Systems Research, 2*(1), 1–14.

Parkhe, A. (1993). Messy research, methodological predispositions and theory development in international joint ventures. *The Academy of Management Review, 18*(2), 227–268.

Patton, M. Q. (1990). *Qualitative evaluation and research methods*. Newbury Park, CA: Sage Publications.

Perry, C. (1998). Processes of case study methodology for postgraduate research in marketing. *European Journal of Marketing, 32*(9/10), 785–802.

Perry, C., Riege, A., & Brown, L. (1999). Realism's role amongst scientific paradigms in marketing research. *Irish Marketing Review, 12*(2), 16–23.

Phillips, E. M., & Pugh, D. S. (2000). *How to get a PhD: Handbook for students and their supervisors* (3rd ed.). Buckingham: Open University Press.

Remenyi, D., & Money, A. (2004). *Research supervision: For supervisors and their students*. Curtis Farm, Kidmore End, UK: Academic Conferences Limited.

Richards, L. (1993). Writing a qualitative thesis or grant application. In K. Beattie (Ed.), *So where's your research profile? A resource book for academics*. South Melbourne, Australia: Union of Australian College Academics.

Robson, C. (1993). *Real world research*. Oxford: Blackwell.

Romano, C. (1989). Research strategies for small business: A case study. *International Small Business Journal, 7*(4), 35–43.

Rubinstein, D. (1981). *Marx and Wittgenstein*. London: Routledge & Kegan Paul.

Saunders, M., Lewis, P., & Thornhill, A. (1997). *Research methods for business students*. London: Pitman Publishing.

Seale, C. (1999). *The quality of qualitative research*. London: Sage Publications.

Silverman, D. (2004). *Interpreting qualitative data: Methods for analysing talk, text and interaction* (2nd ed.). London: Sage Publications.

Simpson, L. (2003). *Retail sales promotion in the supermarket industry: A tri-country comparison of New Zealand, Singapore and Australia*. PhD thesis, University of Otago, Dunedin, New Zealand.

Stoecker, R. (1991). Evaluating and rethinking the case study. *The Sociological Review, 39*(1), 88–112.

Strauss, A. (1987). *Qualitative analysis of social science*. Cambridge: Cambridge University Press.

Summers, J. O. (2001). Guidelines for conducting research and publishing in marketing: From conceptualisation through the review process. *Journal of the Academy of Marketing Science, 29*(4), 405–415.

Ticehurst, G. W., & Veal, A. J. (2000). *Business research methods: A managerial approach*. Frenchs Forest, NSW: Longman.

Trochim, W. M. (2000). *The research methods knowledge base.* Retrieved July 6, 2006, from http://trochim.human.cornell.edu/kb/index.htm

Tsoukas, H. (1989). The validity of idiographic research explanations. *The Academy of Management Review, 14*(4), 551–561.

Whetten, D. A. (1989). What constitutes a theoretical contribution? *The Academy of Management Review, 14*(4), 490–495.

Yin, R. K. (2003). *Case study research: Design and methods* (3rd ed.). Applied Social Research Methods Series, Vol. 5. Thousand Oaks, CA: Sage Publications.

7

Private Label and Manufacturer Brand Research Execution

Overview

This chapter discusses the research execution of the study on the coexistence of private label and manufacturer brands in food product categories in a grocery retail landscape characterised by high retail consolidation and concentration. The main aspects discussed in this chapter are data collection and data analysis procedures. Specifically, the chapter covers data collection, a pilot study, data analysis, the limitations of the methodology and ethical considerations.

Data Collection

Techniques of Data Collection

A number of data collection techniques and sources were considered for this book in order to ensure that the most appropriate ones were chosen. A choice was made from the following list: documents; observation (direct and participant); interviews (Miles and Huberman 1994; Trochim

R. Chimhundu, *Marketing Food Brands*,
https://doi.org/10.1007/978-3-319-75832-9_7

2000; Yin 2003, 2013); archival records and physical artefacts (Yin 2003). These sources are considered to be the most commonly used in case study research (Yin 2003) and are not mutually exclusive. Documentation, archival records, direct observation and in-depth interviews were chosen for this research because of their combined capability to fully address the evidence requirements of this book. The case research therefore combines multiple methods of data collection (Eisenhardt 1989) that are triangulated by the researcher (Yin 2003, 2013) to achieve a more accurate and better understanding of the "whats", "hows" and "whys" of the coexistence of manufacturer brands and private label in FMCG product categories in a highly concentrated marketing environment. The research involved the collection of both secondary and primary data. With respect to some specific research issues, the two types of data complement each other to address the issues. What follows is a further discussion of pertinent issues related to secondary and primary data.

Secondary Data

The book makes use of a preliminary research stage that looks at private label share trends in four developed economies; this background study informs subsequent research. This study largely employs secondary data sources, although it is not solely limited to these. It was necessary to finalise the secondary data collection and analysis stage in this research before the pilot study and the main research stage. The secondary data stage provides important qualitative and quantitative background information (McDaniel and Gates 1998) about manufacturer brands and private label in the categories, enabling more informed investigations in the later stages. In certain instances, it may be "wasteful to collect new information" (Ticehurst and Veal 2000: 46) through primary research where secondary data exist to address the research issues adequately. But at the same time, it was also important to ensure that this case study research, which is set in the realist paradigm, benefited from the triangulation of findings (Miles and Huberman 1994; Yin 2003, 2013). Therefore, the idea of "corroborating findings" (McDaniel and Gates 1998: 76) is considered to be paramount. Documentation and archival records used, inter alia, as sources of secondary data, have enabled this corroboration.

Secondary data sources can be divided into two: namely, the company/companies under research (internal databases) and other external organisations or people (external databases) (McDaniel and Gates 1998). As regards external providers of secondary data, extra care was taken to assess the credibility of such sources of evidence. It is generally the case, though, that well-known commercial/research organisations tend to be more reliable, as the continued existence of such organisations largely depends on the credibility of the data they supply (Saunders et al. 1997: 173). Use was also made in this research, of a comprehensive checklist developed for the evaluation of secondary data; this checklist details the overall suitability, precise suitability and costs and benefits (Saunders et al. 1997: 177) of the data.

In-Store Category Observation Study (Primary Data)

This observation study has gone beyond restricted definitions of observation (e.g. Saunders et al. 1997; Ticehurst and Veal 2000) that seem to imply that such studies are all about observing people. The author took the view that observation studies are broad in nature and can be applied to a wide variety of phenomena (Cooper and Schindler 2001; Gross et al. 1971; Trochim 2000; Yin 2003). Specifically, Cooper and Schindler (2001) divide observations into two broad types: behavioural observation, which involves observing persons, and non-behavioural observation, which involves observing phenomenon like physical situations, records, processes and so on. The observation study for this research therefore fitted into the classification of non-behavioural observation since it examined physical manufacturer brand and private label supermarket category situations related to shelf space, shelf facings, shelf position, number of brands/products, and even product quality and other relevant quantitative as well as qualitative issues.

The quantitative and qualitative information derived from observation was triangulated with data from secondary sources and interviews, thus facilitating the convergence of information (Gross et al. 1971; Yin 2003,

2013). The category observation data also set the stage for discussions at the chosen sites. With regard to chronology, therefore, at store sites observation generally came before interviews, thus facilitating enriched discussions. The same retail sites chosen for interviews were the sites where observation studies took place. A category observation form, developed for this research, was used in the observation study, and a tape measure was also used in the exercise. The category observation study took two to two-and-a-half hours to complete in each of the 18 stores studied. Details of the category observation study undertaken are given in Tables 7.1 and 7.2. The stores covered are anonymised as per confidentiality agreements with research participants and consent forms were signed by the participants. The codes QR and ST, subdivided into Q, R, S and T and then further split into groups Q1 to Q6, R1 to R6, S1 to S3 and T1 to T3, are used to aid anonymisation.

Five forms were completed in each store for each of the food categories of interest—that is, milk, flour, cheese, breakfast cereals and tomato sauce.

Table 7.1 In-store category observation study, retail chain QR

Supermarket chain	Store (and category observation identification code)	Number of forms completed
Supermarket Q	Q1	5
	Q2	5
	Q3	5
	Q4	5
	Q5	5
	Q6	5
		30
Supermarket R	R1	5
	R2	5
	R3	5
	R4	5
	R5	5
	R6	5
		30
Total		*60*

Source: Table created for this book based on the category observation study undertaken

Table 7.2 In-store category observation study, retail chain ST

Supermarket chain	Store (and category observation identification code)	Number of forms completed
Supermarket S	S1	5
	S2	5
	S3	5
		15
Supermarket T	T1	5
	T2	5
	T3	5
		15
Total		*30*

Source: Table created this book based on the category observation study undertaken

In-Depth Interviews (Primary Data)

The in-depth interview technique chosen for this book enabled the researcher to collect "a rich set of data" (Saunders et al. 1997: 215) on the coexistence of manufacturer brands and private label in the selected product categories. Interviews make it possible to dig into and understand the reasons for the decisions taken (Saunders et al. 1997) by FMCG managers, in terms of the intentions of the managers as well as their likely reactions to phenomena that have an impact on their business. In addition, the interview technique enables flexible and responsive interaction between the interviewer and interview participants, allowing questions to be made clear to participants, meanings to be probed and discussion items to be covered from a variety of angles as necessary (Sykes 1991). The research used the semi-structured process of in-depth interviewing as it allows prior theory to be accommodated. The in-depth interviews therefore combined inductive, free-flowing interviewing with deductive, structured interviewing (Carson et al. 2001).

In-depth interview data were corroborated by data from archival records, documentation and observational study to take care of the common problems of "bias, poor recall, and poor or inaccurate articulation" that are sometimes associated with the interview method of data collection (Yin 2003: 92). The idea of triangulating interview data using multiple data sources is therefore important, as it enables testing for convergence (Parkhe 1993). To further ensure the avoidance of interview bias, the researcher did not impose a reference frame on the interview participants, both when asking the questions and when interpreting them (Easterby-Smith et al. 1991). As advised by Easterby-Smith et al. (1991), probes were used in such a way that they did not lead participants in a certain direction. Some of the probing techniques suggested by the same authors (p. 80) were used as necessary. These included basic probes, explanatory probes, focused probes and silent probes, as well as drawing out, giving ideas/suggestions and mirroring/reflecting.

A record of the interview in the form of notes was compiled immediately after each interview (Healey and Rawlinson 1994; Robson 1993; Saunders et al. 1997), mainly because, if this is not done, "the exact nature of explanations provided may be lost as well as general points of value" (Saunders et al. 1997: 224). This also made it possible to take relevant points into the next interview, as well as allowing continuous reflection and analysis during data collection. In addition, the interviews were tape recorded, with permission from the participants. Despite concerns such as the possibility of inhibiting respondents, the tape recording of in-depth interviews is quite common (Ticehurst and Veal 2000) and has positives that arguably outweigh the associated negatives. For this research, the tape recordings were used for two key benefits: namely, to enable the compilation of more detailed research interview data, and to provide a wealth of relevant, accurate quotations.

An interview protocol/checklist was used as an instrument of data collection for the interviews. The items for discussion were based on the specific research issues and resultant data needs (Ticehurst and Veal 2000). The researcher made use of items for discussion rather than set questions because each question was shaped "according to the circumstances of a particular interview" (Ticehurst and Veal 2000: 98). The research was pursuing a line of enquiry, but the questions were fluid and not rigid (Rubin and Rubin 1995). The interview protocol acted as a guide in this regard. It should be noted that all interview items were applicable to retailers, manufacturers and consultants, so it was not necessary to devise multiple versions of the interview protocol. Interviews varied in length, with the longer ones taking slightly over an hour; most lasted around 40 to 45 minutes. However, in some situations the interview time was shared between participants, for instance in one supermarket store where three interviews were held separately with three different participants handling different parts of the interview over an hour or so. The semi-structured nature of the in-depth interviews conducted, due to the existence of prior theory, was such that the interviews were more focused on specific lines of enquiry while at the same time allowing room for flexibility on other lines of enquiry. Little time was therefore wasted on issues that were not relevant to the research. Ticehurst and Veal (2000: 97) have noted that such interviews can take "half an hour or more".

With respect to preparation for and entering the field, the researcher made contact with case study organisations by telephone initially, to briefly introduce the research and establish the right contact. Relevant details of the research followed in the form of a covering letter, information sheet for participants and consent form. A follow-up approach was made soon after, to set up the dates and times of the interviews. Finer details relating to preparation for interviews, opening the interviews and conducting the interviews were based on a checklist (Saunders et al. 1997: 231–232) that is in the literature for collecting primary data. Details of the research interviews undertaken are furnished in Tables 7.3, 7.4, 7.5 and 7.6. The interviews have been anonymised as per the agreement with study participants and study organisations. Of the 46 interviews conducted, 34 were held face-to-face and 12 by telephone using a speaker phone. All but one of the interviews were tape recorded for the later compilation of detailed research interview data.

Table 7.3 Research interviews (retail chain QR)

Operation	Research participant	Mode of interview
Regional company head office	Four senior managers (two interviews as one was combined with three managers)	Face-to-face
Regional company (sister company) head office	One senior manager	Face-to-face
Supermarket chain Q		
Six stores (Q1 to Q6)	Nine managers (eight interviews as one was combined)	Face-to-face
Supermarket chain R		
Six stores (R1 to R6)	Ten managers (ten interviews)	Face-to-face

Source: Table prepared for this book based on research interviews conducted

Table 7.4 Research interviews (retail chain ST)

Operation	Research participant	Mode of interview
Head office	One senior manager	Face-to-face
Supermarket chain S		
Stores S1 to S3	Five managers (five interviews)	Face-to-face
Supermarket chain T		
Stores T1 to T3	Three managers (three interviews)	Face-to-face

Source: Table prepared for this book based on research interviews conducted

Table 7.5 Research interviews (manufacturers/suppliers)[a]

Organisation	Research participant	Mode of interview
Manufacturer W1 (Flour & breakfast cereals)	Brand manager	Face-to-face
Manufacturer W2 (Breakfast cereals)	Marketing manager	Face-to-face
Manufacturer W3a (Milk & cheese)	National business manager (key accounts)	Telephone
Manufacturer W3b (Flour)	Sales director	Telephone
Manufacturer W4 (Breakfast cereals)	Brand manager	Telephone
Manufacturer W5 (Breakfast cereals)	CEO	Face-to-face
Manufacturer W6 (Tomato sauce)	Marketing manager	Telephonic
Manufacturer W7 (Other FMCG)	Marketing manager	Telephone
Manufacturer W8 (Other FMCG)	General manager	Telephone
Manufacturer W9 (Milk)	Business development manager	Telephone
Manufacturer W10 (Cheese)	Managing director	Telephone
Manufacturer W11 (Tomato sauce)	Owner	Telephone
Manufacturer W12 (Flour)	Food division manager	Telephone

Source: Table prepared for this book based on research interviews conducted

[a]Reprinted from *Australasian Marketing Journal*, Vol. 23 No. 1, Chimhundu et al., Manufacturer and retailer brands: Is strategic coexistence the norm? Page 54, Copyright (2015), with permission from Elsevier

Table 7.6 Research interviews (consultants)[a]

Organisation	Research participant	Mode of interview
Consulting company Y1	Director	Face-to-face
Consulting companyY2	Business development manager	Telephone
Consulting company Y3	Managing director	Telephone

Source: Table prepared for this book based on research interviews conducted
[a]Reprinted from *Australasian Marketing Journal*, Vol. 23 No. 1, Chimhundu et al., Manufacturer and retailer brands: Is strategic coexistence the norm? Page 55, Copyright (2015), with permission from Elsevier

Pilot Study

The research made use of a pilot study. The objectives of the pilot study were to assist in the development of "relevant lines of questions" (Yin 2003: 79) and to test the appropriateness of the interview protocol and the in-store category observation form. The selection of these pilot study organisations was largely based on "convenience, access, and geographic proximity" (Yin 2003: 79). The pilot study consisted of five interviews and two category observation exercises, as follows. *Retail chain QR*: one supermarket store, Code Q; joint, face-to-face interview with two managers; trial in-store category observation carried out. *Retail chain ST*: one supermarket store, Code T; face-to-face interview with one manager; trial in-store category observation carried out. *Manufacturers*: food manufacturing company; interview with managing director (by telephone). *Industry analysts*: international marketing research firm; interview with marketing and communications executive (by telephone). *Data house in the food and grocery industry*: interview with industry analyst, by telephone.

The pilot study was used to shape the main study in four ways. Firstly, it confirmed that all the five categories chosen had private label brands in them. Secondly, in the supermarkets, research interviews would come after category observation (wherever possible) in order to allow the shelf data gathered to be part of the discussions. Thirdly, category observation and measurement would be done only once in each store (not twice as initially planned) since shelf space, facings and positions allocated to participating brands and products do not change on a daily basis, but only after a category review every few months. Fourthly, adjustments to the interview protocol and category observation form were made.

Data Analysis

The approach to analysing data for this book was that "the essence of any analysis procedure must be to return to the terms of reference, the conceptual framework and the questions or hypothesis of the research" (Ticehurst and Veal 2000: 100); a stance shared by a number of other researchers (e.g. Brown 1996; Perry 1998a, b; Yin 2003). Therefore, interview data, in-store category observation data and data on private label share trends were systematically analysed against each of the specific research issues, as a way of testing for adequacy. The data analysis procedure adopted therefore involved "examining, categorizing, tabulating, testing, or otherwise recombining" qualitative and quantitative evidence to address the specific research issues (Yin 2003: 109) and conceptual framework of the study developed in the research issues chapter (Chap. 5). By so doing, the analysis of the information did not lose focus on addressing the research problem (Perry 1998b).

The general analytic strategy adopted by the research was along the lines of Yin's (2003) alternative strategy of relying on theoretical propositions. The data analysis framework used consisted of three components: data reduction, data display, and conclusion drawing and verification (Miles and Huberman 1994). Details of the data analysis components and respective analytical tasks are outlined in Table 7.7. Although not all items listed in this table were employed, most were.

The three components, data reduction, data display and conclusion drawing/verification were treated as being interwoven throughout the entire research process, that is, before, during and after the collection of data (Miles and Huberman 1994). However, the bulk of the analysis of data was done after collection. Although the three components of data analysis are applicable to secondary data, category observation data and in-depth interview data, not all items listed in the "Analytical tasks involved" column of Table 7.7 are applicable to all three. In addition to

Table 7.7 Components of data analysis

Component	General process	Analytical tasks involved	Level of analytical abstraction
Data reduction	The process of selecting, focusing, simplifying, abstracting and transforming the data that appear in written-up field notes or transcriptions.	Tasks include: - Writing summaries - Writing analytical notes - Coding - Identifying themes - Making clusters - Making partitions - Memos	Summarising and packaging the data
Data display	The organised, compressed assembly of information that enables one to see what is happening, and permits conclusion drawing and verification.	Use of: - Matrices - Charts - Networks - Searching for relationships	Repackaging and aggregating the data
Conclusion drawing and verification	The process of attaching meaning to the information and giving confirmation.	Noting and confirming the following: - Regularities - Patterns - Explanations - Possible configurations - Causal flows - Propositions	Developing and testing propositions to construct an explanatory framework

Source: Developed from Miles and Huberman (1994) and Carney (1990) for this book

the general data analysis framework shown in Table 7.7, additional issues related to the analysis of secondary data, in-store category observation study data and in-depth interview data for this research are explained next.

Secondary Data

Quantitative data on long-term private label share trends were subjected to time series analysis using a common algorithm, the moving average

(Saunders et al. 1997). Graphs were used to display the data, thereby revealing patterns that were then interpreted. This part of the study formed part of the preliminary research stage of the book. The preliminary study has made a contribution to the marketing academic literature, since a journal article based on it (Chimhundu et al. 2011) has been published in a prestigious peer-reviewed journal.

In-store Category Observation Data

The data collected on shelf space, shelf facings, shelf position, and number of brands and products were largely quantitative in nature, although there are some qualitative aspects. The data were compiled per category and per store and displayed in matrix and tabular form. Comparisons between manufacturer brands and private label were then made. Both within-case and cross-case comparisons were made. The actual shelf data tables and condensed tables, however, are not shown in this book for confidentiality reasons.

Interview Data

The in-depth interview data collected were largely qualitative in nature. While interview notes were compiled and key themes identified and refined throughout the data collection process, the main part of the analysis came after data collection. The main part of the interview data analysis was approached as follows.

The researcher compiled detailed research interview data from the interview tapes and notes. The researcher did their own transcription in order to capitalise on the advantage of gaining familiarity with the data, as suggested by Gibbs (2007: 15).

> It gives you a chance to start the data analysis. Careful listening to tapes and reading and checking of the transcript you have produced means that you become very familiar with the content. Inevitably you start to generate new ideas about the content.

Ultimately, the researcher was interested in the meanings of the data (Arksey and Knight 1999). Although detailed research interview data were produced, this was done to the level appropriate for the requirements of this research, as transcription can be done at a number of levels (Arksey and Knight 1999; Gibbs 2007; Silverman 2004a, b). It was therefore not necessary to do it to the level required by "linguistic researchers and those interested in discourse analysis" (Arksey and Knight 1999: 141). While every relevant detail of the interviews was captured in the written account, things like digressions and superfluous material were left out. Moreover, while the research interview data are abundant with verbatim quotations from participants, there are parts that were reconstructed as interview notes, which was done in such a way that meaning was not lost. In addition, it is noted that for studies that make use of mixed methods and where an interview guide is employed, with "topic areas pre-specified on an interview guide but the researcher [varying] the wording or order of questions depending on the participant", the research interview data produced may not necessarily have to follow the strict verbatim transcript approach (Halcomb and Davidson 2006: 39). This research has made use of in-store category observation data and private label share trend data in combination with the qualitative study, and although it is largely qualitative in nature, it includes an element of mixed methods research.

The research interview data were coded. Codes were derived from prior theory (and from the interviews) as the research was pursuing certain lines of enquiry in the context of relevant themes. Thematic coding (Gibbs 2007; Krippendorff 2004; Trochim 2000) was therefore judged to be appropriate for this research. A list of codes was compiled (see Appendix) to make it easier to manage the large volumes of data. A hard copy of the research interview data was printed out and coded using a highlighter, with the relevant codes written in the margin with a pen. In a number of instances, particular texts were actually assigned two or more codes where there was overlap. Sections with similar codes were then grouped together on the soft copy of the Microsoft Word document and analysed for meanings and patterns. A template of the research issues was also created in MS Word and all coded data relevant to each specific research issue were grouped under each respective issue and further analysed for meanings and patterns. Relevant tables were created addressing the specific research issues. Quotations that would be used as text evidence were selected. It should be noted that computer packages such as NUD*IST were not employed in this process due to the fact that

> they are not essential for realism research because realism researchers do not need to map all the details of the interviewee's subjective reality, they merely look through some parts of the reality at an external reality and manual coding of interviews can be adequate for this process. (Perry et al. 1999: 19)

Repackaged interview data therefore largely appeared in the form of tables and matrices. Both within-case and cross-case analysis was conducted. While within-case analysis mainly involved the coding of information (labelling data), memoing (theorising ideas about codes and their relationships), analysing meanings and patterns and developing propositions, cross-case analysis looked for similarities and differences between patterns (Miles and Huberman 1994; Perry et al. 1999). Replication logic was employed in cross-case comparisons. Quotations from interviews were used to support theoretical points.

Triangulation of Evidence

Triangulation of secondary data (from documentation and archival records), in-store category observational data and in-depth interview data is key to the outcomes of this research book. An attempt was made to employ most of the four types of triangulation identified by researchers—in other words, data, methodological, investigator and theory triangulation (Denzin 1989; Patton 1987). Data triangulation and methodological triangulation were achieved through the use of different sources of information and methods of data collection (in-depth interviews, category observation, documentation and archival records). Furthermore, research participants were given a chance to validate research interview data soon after its compilation. Theory triangulation was achieved through the deliberate use of multiple perspectives in approaching empirical findings. Furthermore, emergent theories were compared with the extant literature.

Limitations of Methodology

This section summarises what critics of case study research consider to be potential weaknesses of the methodology and discusses how this research dealt with the respective limitations. A number of authors have contributed to the discussion on the arguments against case research (e.g. Bonoma 1985; Firestone 1993; Parkhe 1993; Silverman 2004a, b; Yin 2003, 2013). Their contributions suggest that their main concerns about such research are that it lacks rigour and has little basis for scientific generalisability. Other concerns include the perception that the case study methodology takes too long (often resulting in a lengthy narrative), and the idea that case study research is only appropriate for the exploratory stages of quantitative research projects.

With regard to the issue of apparent lack of rigour, case studies have been associated with anecdotalism, investigator bias (the selection of facts to fit preconceived positions), inadequate documentation and a general lack of systematic approach. This book has adopted an approach that is considered in the literature (Yin 2003) to be rigorous and systematic. It includes making use of prior theory, using multiple sources of evidence, establishing a chain of evidence, allowing interview participants to review interview data/transcripts, addressing rival explanations, using replication logic, using a case study protocol and developing a case study database. In addition, systematic analysis of case study evidence largely based on Miles and Huberman (1994) and Gibbs (2007) was employed. These measures, among others, have brought in a good measure of validity and reliability to the research.

On the issue of there being little basis for scientific generalisability, Yin (2003, 2013) has noted that the question of how one could generalise from a single case can partly be taken care of by the multiple-case design, which should be regarded as "a multiple set of experiments" (p. 10). This research adopts a multiple-case design. Researchers have identified three types of scientific generalisation as sample to population extrapolation (statistical generalisation), analytic generalisation and case-to-case transfer; analytic generalisation and case-to-case transfer are considered to be applicable to case study research (Firestone 1993). Some researchers have warned that it is a misconception to expect statistical generalisation from case research. They argue that case studies "are generalisable to theoretical propositions and not to populations or universes [...] in doing a case study, your goal will be to expand and generalise theories (analytic generalisation) and not to enumerate frequencies (statistical generalisation)" (Yin 2003: 10). This book seeks analytic generalisation and not statistical generalisation.

The concern about case research taking too long and resulting in huge documents is normally associated with ethnographic research (Yin 2003). This book is not set in the ethnographic mode. The last criticism, that case study research is only appropriate for the exploratory stages of bigger, quantitative research projects, can be regarded as a misconception because case study research is a full methodology in its own right, which has been and can be used successfully in exploratory, descriptive or explanatory studies (Yin 2003, 2013).

Ethical Considerations

Ethics in research refers to the appropriateness of the researcher's behaviour in relation to the rights of people (and organisations) who become the subject of the researcher's work or are affected by the research in one way or another (Saunders et al. 1997). It is important to observe research ethics because unethical behaviour can adversely affect research participants, the researcher and the results of the research (Patton 1990). A number of authors have devoted either whole sections or chapters to ethical considerations in a way that underlines the importance of proactive behaviour in dealing with ethics in research (e.g. Miles and Huberman 1994; Saunders et al. 1997; Trochim 2000; Wells 1994). This book has derived its ethical guidelines from a combination of sources that include Saunders et al. (1997), Trochim (2000) and others.

This book observes ethical issues at all stages, that is, the design and initial access stage, data collection stage, and analysis and reporting stage, as advised by Saunders et al. (1997). In summary, the key ethical issues considered include: ensuring voluntary participation and informed consent (Trochim 2000); being honest with participants about why the research is being undertaken and how the data are going to be used;

respecting prospective and current participants' rights to privacy; maintaining objectivity during data collection, analysis and reporting; respecting assurances about the confidentiality of data; respecting assurances about anonymity; and considering the collective interests of participants in the way the researcher uses the data provided (Saunders et al. 1997). The "code of behaviour appropriate to academics and the conduct of research" (Wells 1994, p, 284) was accessed and used by the author to compile an application for ethics approval for this research, which was granted.

Chapter Recap

This chapter has discussed the execution of the research that forms the basis of this book. The chapter has specifically addressed data collection, pilot study, data analysis, limitations of the methodology and ethical considerations. Chapter 8 addresses empirical evidence on the coexistence of private label and manufacturer brands.

References

Arksey, H., & Knight, P. (1999). *Interviewing for social scientists: An introductory resource with examples*. London: Sage.

Bonoma, T. V. (1985). Case research in marketing: Opportunities, problems and a process. *Journal of Marketing Research, 22*(2), 199–208.

Brown, R. (1996). *Key skills for publishing research articles*. Brisbane: Write Way Consulting.

Carney, T. F. (1990). *Collaborative inquiry methodology*. Windsor, Canada: Division of Instructional Development, University of Windsor.

Carson, D., Gilmore, A., Perry, C., & Gronhaug, K. (2001). *Qualitative marketing research*. London: Sage Publications.

Chimhundu, R., Hamlin, R. P., & McNeill, L. (2011). Retailer brand share statistics in four developed economies from 1992 to 2005: Some observations and implications. *British Food Journal, 113*(3), 391–403.

Cooper, D. R., & Schindler, P. S. (2001). *Business research methods* (7th ed.). New York, NY: McGraw-Hill/Irwin.

Denzin, N. K. (1989). *The research act* (3rd ed.). Englewood Cliffs, NJ: Prentice Hall.
Easterby-Smith, M., Thorpe, R., & Lowe, A. (1991). *Management research: An Introduction*. London: Sage Publications.
Eisenhardt, K. M. (1989). Building theories from case study research. *The Academy of Management Review, 14*(4), 532–550.
Firestone, W. A. (1993). Alternative arguments for generalising from data as applied to qualitative research. *Education Researcher, 22*(4), 16–23.
Gibbs, G. R. (2007). *Analyzing qualitative data*. London: Sage.
Gross, N. C., Giacquinta, J. B., & Bernstein, M. (1971). *Implementing organizational innovations*. New York, NY: Basic Books.
Halcomb, E. J., & Davidson, P. M. (2006). Is verbatim transcription of interview data always necessary? *Applied Nursing Research, 19*(1), 38–42.
Healey, M. J., & Rawlinson, M. B. (1994). Interviewing techniques in business and management research. In V. J. Wass & P. E. Wells (Eds.), *Principles and practice in business and management research* (pp. 123–146). Aldershot: Dartmouth.
Krippendorff, K. (2004). *Content analysis: An introduction to its methods*. Thousand Oaks, CA: Sage.
McDaniel, C., Jr., & Gates, R. (1998). *Marketing research essentials* (2nd ed.). Cincinnati, OH: South-Western.
Miles, M. B., & Huberman, A. M. (1994). *Qualitative data analysis: An expanded sourcebook*. Thousand Oaks, CA: Sage Publications.
Parkhe, A. (1993). Messy research, methodological predispositions and theory development in international joint ventures. *The Academy of Management Review, 18*(2), 227–268.
Patton, M. Q. (1987). *How to use qualitative methods in evaluation*. Newbury Park, CA: Sage Publications.
Patton, M. Q. (1990). *Qualitative evaluation and research methods*. Newbury Park, CA: Sage Publications.
Perry, C. (1998a). Processes of case study methodology for postgraduate research in marketing. *European Journal of Marketing, 32*(9/10), 785–802.
Perry, C. (1998b). A structured approach for presenting theses. *Australasian Marketing Journal, 6*(1), 63–85.
Perry, C., Riege, A., & Brown, L. (1999). Realism's role amongst scientific paradigms in marketing research. *Irish Marketing Review, 12*(2), 16–23.
Robson, C. (1993). *Real world research*. Oxford: Blackwell.
Rubin, H. J., & Rubin, I. S. (1995). *Qualitative interviewing: The art of hearing data*. Thousand Oaks, CA: Sage Publications.

Saunders, M., Lewis, P., & Thornhill, A. (1997). *Research methods for business students*. London: Pitman Publishing.

Silverman, D. (2004a). *Interpreting qualitative data: Methods for analysing talk, text and interaction* (2nd ed.). London: Sage Publications.

Silverman, G. (2004b, November 16). Retailers pack new punch in battle with the brands. *Financial Times*.

Sykes, W. (1991). Taking stock: Issues from the literature in validity and reliability in qualitative research. *Journal of Market Research Society, 33*(1), 3–12.

Ticehurst, G. W., & Veal, A. J. (2000). *Business research methods: A managerial approach*. Frenchs Forest, NSW: Longman.

Trochim, W. M. (2000). *The research methods knowledge base*. Retrieved July 6, 2006, from http://trochim.human.cornell.edu/kb/index.htm

Wells, P. (1994). Ethics in business and management research. In V. J. Wass & P. E. Wells (Eds.), *Principles and practice in business and management research* (pp. 277–298). Aldershot: Dartmouth.

Yin, R. K. (2003). *Case study research: Design and methods* (3rd ed.). Applied Social Research Methods Series, Vol. 5. Thousand Oaks, CA: Sage Publications.

Yin, R. K. (2013). *Case study research: Design and methods* (5th ed.). Thousand Oaks, CA: Sage Publications.

8

Empirical Evidence on the Coexistence of Private Label and Manufacturer Brands

Overview

This chapter reports the results of the intensive study carried out on private label and manufacturer brand coexistence in the consumer goods industry. Specifically, it includes: a restatement of the research issues; case summaries; discussions on the balance between private label and manufacturer brands, and balance in a highly concentrated grocery retail landscape; comparative capacity for production innovation and category support; the role of consumer choice; the nature of coexistence between private label and manufacturer brands, and the role of power in that coexistence.

A Restatement of the Research Issues

The primary research question of this book is: *How do manufacturer brands and private label coexist in FMCG/supermarket product categories in a grocery retail landscape characterised by high retail concentration, and how relevant is power to this coexistence?*

R. Chimhundu, *Marketing Food Brands*,
https://doi.org/10.1007/978-3-319-75832-9_8

The primary research question has been decomposed into three subsidiary research questions, and these are: Does a grocery retail environment characterised by high retail concentration lead to an overdominance of private label in relation to manufacturer brands in FMCG/supermarket product categories? How important are aspects of strategic dependency between manufacturer brands and private label in determining the nature of coexistence between the two types of brands in FMCG/supermarket product categories? In an FMCG/supermarket landscape characterised by high retail concentration, and direct competition between brands owned and managed by owners of the grocery retail shelves and those owned and managed by their suppliers, what is the role of power in the coexistence relationship between the two types of brands in the product categories?

These research questions have given rise to a series of research issues that can be divided into three major themes, namely: the balance between manufacturer brands and private label in the categories (research issues 1, 2a, 2b and 2c); innovation, category marketing support and consumer choice (research issues 3a, 3b, 4a, 4b, 4c and 5); and category strategic policies on the coexistence of private label and manufacturer brands (research issues 6 and 7). The research issues are outlined in the following section.

Balance between Manufacturer Brands and Private Label in FMCG Product Categories

Research Issue 1: What is the general nature of the state of balance between private label and manufacturer brands in FMCG product categories?

Research Issue 2a: In a grocery retail landscape characterised by high retail concentration, what is the nature of the state of balance between private label and manufacturer brands in FMCG product categories?

Research Issue 2b: How does the balance between private label and manufacturer brands in FMCG product categories compare by category?

Research Issue 2c: How does the balance between private label and manufacturer brands in FMCG product categories compare between grocery retailers?

Product Innovation, Category Marketing Support and Consumer Choice

Research Issue 3a: How do private label and manufacturer brands compare on capacity to innovate in the FMCG product categories?

Research Issue 3b: How do private label and manufacturer brands compare on capacity to contribute to category marketing support and development?

Research Issue 4a: What is the state of awareness of FMCG retailers on the comparative capacity of private label and manufacturer brands to innovate and give marketing support to the product categories?

Research Issue 4b: What is the strategic stance of the FMCG retailers with respect to this comparative capacity to innovate and give marketing support to the product categories?

Research Issue 4c: How is this strategic stance related to policies on the coexistence of private label and manufacturer brands in FMCG product categories?

Research Issue 5: What role is played by consumer choice in shaping the coexistence of private label and manufacturer brands in FMCG product categories?

Category Strategic Policies on the Coexistence of Private Label and Manufacturer Brands in FMCG Product Categories

Research Issue 6: What is the nature of the coexistence relationship between private label and manufacturer brands in FMCG product categories and how is it driven?

Research Issue 7: What role is played by power in the mode of coexistence between private label and manufacturer brands in FMCG product categories?

Case Summaries: FMCG Categories and Grocery Retailers

The information that is presented in this section is derived from a number of sources that include websites, company reports and research interviews. Some of the information, however, has been withheld to preserve the anonymity of contributors. This is particularly so in the case of specific information about the FMCG retail chains. Most of the information on the product categories is derived from interviews carried out within the retail groups, anonymised for instance as QR1, QR2, ST1 and so on.

FMCG/Supermarket Categories

Milk Category

Within the supermarkets, milk and milk products are grouped into a number of different categories and aisles such as standard milk, cream, flavoured milk and long-life milk. This study largely focuses on standard milk/fresh milk, although in the process, the other subcategories are made reference to. The standard milk category is a fairly commoditised category. Despite this, a considerable amount of product innovation is taking place within it. While private label would generally be more profitable to retailers than manufacturer brands, margins in the milk category have largely been eroded by the competitive activity between the rival grocery retail chains. As far as the management of this category is concerned, both the retailers and the manufacturers have an input.

Flour Category

The flour category incorporates different types of products that include high-grade, self-raising, standard, wholemeal and gluten-free flour. The category is also a largely commoditised one. Again, a considerable amount of product innovation is taking place. From a profit point of view, private label brands generally offer the retailers higher margins than manufacturer brands. Moreover, in managing this category, both retailers and manufacturers have an input.

Cheese Category

Cheese can be divided into standard cheese, speciality cheese and cultured products, and within the supermarket, these are separate categories. The study focuses on standard cheese. Products that can be found in the standard cheese category include block cheese, grated cheese and sliced cheese. There is a higher level of innovation in the speciality segment, although there is still a considerable amount of innovation in the standard cheese category. As far as commoditisation is concerned, it depends on the segment; for instance, standard cheese tends to be more inclined towards commoditisation. The segment is also generally price driven, with whatever is on promotion selling more. In the management of this category, there is input from both manufacturers and retailers.

Breakfast Cereals Category

The breakfast cereals category consists of a relatively high number of subcategories that include cornflakes, muesli, wheat biscuits, oats, crackers, and branded cereals such as Coco Pops and Rice Krispies. The category is a non-commoditised one and a good deal of product innovation is taking place. In certain subcategories such as cornflakes, technological sophistication is much higher than, say, in the milk and flour categories. Generally, private label brands have higher margins than manufacturer brands in this category, and in the management of the category, there is input from both the manufacturers and the retailers.

Tomato Sauce Category

This category belongs to the larger product group of sauces. The tomato sauce category can be divided into canned and bottled. The category is non-commoditised and there is a considerable degree of innovation taking place around the product and packaging. In terms of profit margins, private label brands tend to return higher margins for the retailers than manufacturer brands. Furthermore, with regard to management of the category, both the manufacturers and the retailers have an input.

Grocery Retail Chains and Selected Supermarket Groups

The New Zealand grocery retail industry is highly concentrated by world standards. It is largely a duopoly. Sources place the two-firm concentration ratio of this industry as being between 95 and 98%. The two umbrella retail chains are both giants in this industry but they are different in terms of their structures. One is a decentralised organisation that is run as a cooperative and has regional companies (regional cooperatives) and owner-operated stores. For this retail chain, the main focus of this research is on one of its regional companies. The other retail chain is run as a pure retail chain that is centralised. Therefore, there are differences in the levels of autonomy between the two grocery retail chains. There is a higher level of autonomy within the operating units of the former (including store level autonomy) than there is within the latter. The two retail chains have been anonymised as QR and ST. Both retail organisations have branded supermarket chains within them and two from each were selected for this research (anonymised as Q and R for one retail organisation, and S and T for the other).

Balance Between Private Label and Manufacturer Brands

Analysis of Research Issue 1

Research Issue 1: What is the general nature of the state of balance between private label and manufacturer brands in FMCG product categories?

A preliminary study was undertaken that sought to examine private label and manufacturer brand long-term share trends, and to establish the relationship of these trends to FMCG retail concentration and to retailer category strategy on manufacturer brands and private label. The study made use of statistical data on aggregate private label share trends as well as other data from four developed economies: the UK, the USA, Australia and New Zealand. The grocery retail concentrations of these four countries are: the UK, 65% and the USA, 36% (ACNielsen 2005), with both figures being five-firm concentration ratios; Australia, 74% with a two-firm concentration ratio (Anonymous 2005); and New Zealand, 98%, also with a two-firm concentration ratio (ACNielsen 2005). Data collected on private label share trends from these four economies covered a period of 14 years and were analysed using time series analysis (moving average). The results indicate that there are long-term equilibrium points between private label and manufacturer brands in grocery product categories (Chimhundu et al. 2011).

As an additional dimension, a further study introduced quality tiers of private label into the examination of private label–manufacturer brand coexistence. It investigated to what extent the different quality tiers of private label can affect the general nature of the equilibrium between private label and manufacturer brands. The same time-series trends in the four countries were analysed against the retail concentration and private label quality spectrum in the four countries. It was found that there are two states of equilibrium between private label and manufacturer brands, and these are premature and mature equilibrium. Premature equilibrium is a lower level equilibrium where private label share stabilises at a lower level. This happens when only low-quality tiers of private label brands are used and/or when the grocery retail industry has a low level of consolidation/concentration. Mature equilibrium is the stabilisation of private label share at a higher level, and this happens when full exploitation of the private label spectrum is employed in a high-concentration grocery retail environment (Chimhundu and Chadee 2013).

With respect to research issue 1, the following research propositions can be advanced:

Research Proposition 1a: Equilibrium points exist in the aggregate long-term share trends of private label and manufacturer brands in FMCG product categories.

Research Proposition 1b: There are two possible states of equilibrium, premature and mature equilibrium, in the long-term share trends of private label and manufacturer brands in FMCG product categories.

These findings were taken as a foundation, and it was necessary to go a step further by conducting a more detailed analysis of the coexistence of the two types of brands in one of the four markets, using subsequent research issues. The FMCG/supermarket industry is that of New Zealand, and the industry has a very high retail consolidation and concentration as has already been indicated. In analysing the research issues from here onwards, research interview quotations[1] are used, in addition to other research evidence.

Balance in a Highly Concentrated Grocery Retail Landscape

Analysis of Research Issue 2a

Research Issue 2a: In a grocery retail landscape characterised by high retail concentration, what is the nature of the state of balance between private label and manufacturer brands in FMCG product categories?

High retail concentration naturally gives retailers power over manufacturers. From a profit point of view, it was also established that private label brands generate higher margins for the retailers than manufacturer brands. Under such circumstances, it is reasonable to expect the retailers to want to implement a strong share growth regime for their private label.

[1] *Some of the research interview quotations in this chapter were previously published in the following journals:International Journal of Business and Management, Vol. 5 No. 9, Pages 13, 14, 15 and 16 (Chimhundu et al. 2010), Copyright Canadian Centre of Science and Education 2010.*
International Journal of Marketing Studies, Vol. 4 No. 6, Page 41 (Chimhundu 2012), Copyright Canadian Centre of Science and Education 2012.
Asia Pacific Journal of Marketing and Logistics, Vol. 27 No. 3, Pages 374, 375, 376 and 377 (Chimhundu et al. 2015), Copyright Emerald 2015.
Australasian Marketing Journal, Vol. 23, Pages 54, 56 and 57 (Chimhundu et al. 2015), Copyright Elsevier 2015.
Journal of Brand Management, Vol. 23 No. 5, Pages 32, 33 and 35 (Chimhundu 2016), Copyright Macmillan Publishers 2016.

The question of equilibrium between private label and manufacturer brands in grocery retail categories in New Zealand is looked at in this section from an overall retail chain perspective. The main discussions pertaining to this topic were conducted with senior level managers. A total of seven senior managers provided information on this issue.

A senior manager at retail chain QR said: "Overall, we [the grocery retail chain] are currently at around about 12.6[%] dollar share nationally in all categories [...] and ideally we would like to get it up to around 15[%]" (Interview QR2). Another senior manager stated: "Private label in the [region] overall has about 14% share of total private label business. In [this regional company] it's about 13.8%" (Interview QR3).

This state of affairs shows that there is no big difference between the aggregate share situation established by the preliminary study (which took into account data only up to 2005) and this retail chain's situation at the time of the main study, when the interviews were conducted with the managers. There has not been a significant increase in the size of the private label (in relation to the manufacturer brand) as might be expected in an environment characterised by high retail consolidation and concentration. There has, however, been a share rise of a limited nature.

On the same issue, the situation at the other grocery retail chain (characterised as ST) is summarised in the statement: "At the moment we are only about 15% [private label share]" (Interview S3). Again, this situation is not radically different from the earlier, aggregate share results of the preliminary study. It is not too different either from the share situation of the other retail chain, QR. Therefore, these results of the main study for the New Zealand grocery retail industry are in line with the findings of the preliminary study on private label trends.

The discussion was taken a step further to establish the expectations of the retail chains as far as the balance between private label and manufacturer brands is concerned, and the rationale associated with such a balance.

With respect to grocery retail chain QR, two senior managers gave comments. In Interview QR2, one said:

> It is difficult for us to go from 12.6% dollar share to 15%; it's a significant increase. We are in most of the categories in the stores, so it's difficult for us to get another point one of a share point, let alone another three share points [...]. That's by no means an easy task. (Interview QR2)

In Interview Q1, the other senior manager said: "What we have stated as a company is that we want to increase private label by 0.5% share every year for the next five years." The manager added: "Depending on the category, if you are looking at overall, I would say anything that is above 20% would have an effect on the category or an effect on the total business." These three comments actually indicate modest expectations and ambitions on the part of the private label in its coexistence with the manufacturer brand in this retail chain. There does not appear to be a burning desire for, or aggressive stance towards, radically boosting private label share in relation to manufacturer brands. Delving into some of the reasons for these modest expectations and ambitions, the discussions yielded comments related to specific factors, some of which are investigated in-depth in the analysis of later research issues. These factors include the role of innovation in the categories, the role of consumer choice and the small size of the New Zealand market.

With regard to the statement that "anything that is above 20% would have an effect on the category or effect on the total business" (Interview QR1), this research participant further indicated that the negative impact of private label dominance would in fact be on "innovation and spend. If private label becomes too big, then the proprietary brand will withdraw from marketing, innovation [...]; and that's where it [innovation] all comes from because private labels aren't innovators." This is a recognition that the need for product innovation in the categories by manufacturer brands is an integral part of how manufacturer brands and private label should be balanced in the categories. While withdrawal from marketing and innovation is one aspect, another aspect related to it in the New Zealand context is supply:

> If we do that strategy [private label dominance] over here, we would probably end up shutting down most of the manufacturers in New Zealand because if they can't support their own brand, they can't exist to support private label. It's very much a balance. We will have to work in harmony. (Interview QR2)

In addition, consumer choice is seen as an important consideration in balancing the two types of brands, and such choice is seen as not being achievable through private label dominance: "Consumers want to come into our stores and have choice [...]. They want private label offering and branded players" (Interview QR2).

Furthermore, the small size of the New Zealand market plays a part in the determination of the appropriate balance between private label manufacturer brands:

> The New Zealand market is too small for private label to have dominance overall in the categories and across all categories in the stores. There would always need to be a balance between private label and proprietor brands, so we would never want to get to a situation where the private label is the dominant player in all the categories. (Interview QR2)

While issues related to innovation (and the need to take care of suppliers in the categories), consumer choice and the small size of the New Zealand market seem to have a stake in the balancing of manufacturer brands and private label in retail chain QR, as discussed, it is also worth having a look at the situation regarding these issues in retail chain ST for comparison purposes.

An earlier statement established the current state within retail chain ST: "At the moment we are only about 15% [private label share]" (Interview S3); further comments relating to the retail chain's private label expectations and ambitions are given below. It should be noted, however, that this particular comment was made at store/branch level:

> Our target is to get 60% in private label sales [...] overall [...]. In some stores it's more [than 15%]. We realistically think that we can have 40% easily and at a push, we can get to some of the European and American stores where they have got 50% private label. (Interview S3)

While the high figures indicated could not be triangulated with data provided in other interviews in the same retail chain (ST), the comment reflects a more ambitious and aggressive approach to private label. How far this goes could not be established with certainty, but what seems to be supported by the other interviews is that this retail chain is more ambitious than the other one in relation to its private label plans. This could be taken as an area of difference between the two grocery retail chains in New Zealand, with one (retail chain QR) having modest ambitions and the other (retail chain ST) having more than modest ambitions. The differences in retailer strategic objectives in this regard are reflected in this quotation:

> Some retailers may want to drive their private label share as high as possible and other retailers are saying […] "we don't want to go all the way" […] because they have a different philosophy and different strategy, and may talk about leaving it to the supplier to bring innovation. And again, the innovation of the supplier flows through to private label unless you drive a business that is a pure private label business only […]. (Interview W3a)

What is important to note, though, is that even with the retail chain that has bigger private label dominance ambitions, there is still an acknowledgement of the necessity for private label and manufacturer brands to coexist. The difference, therefore, lies only in the perception of where the ultimate equilibrium should lie. As stated by a senior manager at the head office of the ST chain, "We certainly don't see private label being entirely dominant and being the only offer in the categories" (Interview ST1). Factors such as the need for innovation in the categories, the need to take care of suppliers, the small size of the New Zealand market and consumer choice were expressed within both retail chain QR and retail chain ST, as is discussed next.

Commenting on why coexistence is important and why private label dominance is suboptimal, one manager had this to say: "I guess we need to continue to have total category innovation so that they supply […] private label" (Interview ST1). The manager noted the importance of maintaining an equilibrium point the retail chain must not go beyond, otherwise "The point comes where suppliers stop innovating" (Interview ST1). Very high retail consolidation and concentration would not necessarily lead to private label overdominance:

> I guess New Zealand is such a small market […]. We see a lot of room for movement in terms of private label to grow […] as an organisation, but we also see the need and balance, for manufacturer brands to continue to help us drive innovation in the categories. So, private label will be a key differentiation factor versus our competitors; versus also the manufacturer brands. (Interview ST1)

On the same issue of not going beyond a certain percentage as far as private label share/penetration is concerned, this manager gave the justification that "You still have got to look after your suppliers because at the end of the day, they have got a place in the marketplace as well, and it's to no one's benefit to get rid of the suppliers" (Interview S3). In Interview ST1, in addition, the importance of consumer choice in the equation was expressed.

The evidence provided in this discussion under research proposition 2a supports the existence of equilibrium between manufacturer brands and private label; Part of this work has appeared previously (Chimhundu et al. 2015). This topic is looked at, in this discussion, from the viewpoint of the very high retail concentration in New Zealand not having translated itself into a radical share increase on the part of private label. Despite the private label ambitions of the two retail chains, which are both geared towards increasing private label penetration from current levels, there is still an end in sight at both retail chains; a point beyond which the private label will not go, either by way of being unable to or not wanting to go, and this brings about the ultimate equilibrium. Therefore, with respect to research issue 2a (In a grocery retail landscape characterised by high retail concentration, what is the nature of the state of balance between private label and manufacturer brands in FMCG product categories?), the discussion in this section gives rise to the following research proposition:

Research Proposition 2a: In a grocery retail landscape characterised by high retail concentration, there is an existence of equilibrium points between private label and manufacturer brands in FMCG product categories.

Analysis of Research Issues 2b and 2c

Research Issue 2b: How does the balance between private label and manufacturer brands in FMCG product categories compare by category?

Research Issue 2c: How does the balance between private label and manufacturer brands in FMCG product categories compare between grocery retailers?

This section is based on the results of the category observation exercises that were carried out. The tables with the data have not been included in this book for confidentiality reasons. A comparison of the categories studied shows that there are differences in private label penetration in these categories. Categories such as milk and flour have high private label penetration, while categories such as breakfast cereals and tomato sauce have relatively low private label penetration. The differences between categories were echoed in the interviews in both retail chains. At retail chain QR, "When you start splitting it down from category to category, it differs significantly" (Interview QR2). And at retail chain ST,

> Where it's a lot of highly commoditised categories, we would want to increase share. Sometimes in highly commoditised categories, it's not our intent to increase our share to over the dominant position. (Interview ST1)

Therefore, contributing to the aggregate private label share discussed earlier are different degrees of balance between manufacturer brands and private label in the categories. These differences are also reflected in a variety of other aspects studied, such as shelf space and facings, shelf position and number of brands and products in the categories. In relation to research issue 2b (How does the balance between private label and manufacturer in FMCG product categories compare by category?), the discussion in this section gives rise to the following research proposition:

Research Proposition 2b: The balance between private label and manufacturer brands in FMCG product categories differs from category to category.

Furthermore, while there are differences between categories, it was also found that the situation within the supermarket chains Q and R (grocery retail chain QR) and S and T (grocery retail chain ST) is not the same, whether one analyses it by grocery retail chain or by supermarket chain. Each was found to have measures that were different in relation to the others. However, between the two grocery retail chains and between their individual supermarket groups, there is a similar pattern where similar categories have high levels of private label penetration (e.g. milk and flour in both retail chains QR and ST) and low levels of private label penetration (e.g. breakfast cereals and tomato sauce). In this respect, regarding research issue 2c (How does the balance between private label and manufacturer brands in FMCG product categories compare between grocery retailers?), the discussion in this section gives rise to the following research proposition:

Research Proposition 2c: The balance between private label and manufacturer brands in FMCG product categories differs between grocery retailers but follows a similar pattern across similar categories of the different grocery retailers.

In addition, it is only in a limited number of situations that the private label is seen as being slightly overexposed. There are some situations in which the manufacturer brand has been overexposed as well. From a merchandising perspective, therefore, the private label has not dominated the manufacturer brand. In most categories, manufacturer brands still dominate. Moreover, a strategy that was found to be common within all five main categories studied and all four supermarket groups was the practice of displaying private label close to leading manufacturer brands. A further discussion of these aspects of coexistence will be conducted in the exploration of later research issues that deal with strategic issues related to the management of the categories.

Comparative Capacity for Product Innovation and Category Support

Analysis of Research Issue 3a

Research Issue 3a: How do private label and manufacturer brands compare on capacity to innovate in the FMCG product categories?

This research issue is addressed from two complementary points of view: capacity for product innovation and rate of product innovation.

With regard to capacity for innovation, Table 8.1 was prepared from research interviews and other information collected from company websites. A description of resource and expertise requirements for innovation was established in the interview process, particularly in an interview carried out with one consultant, and then based on the information collected via interviews, websites and documentation, an assessment of each respective retail and manufacturer organisation was made to determine the organisation's capacity to undertake product innovation related activities in house. The R&D resources and expertise requirements that were taken into account include product development laboratories, pilot plant, computer software, food labelling software, and technical staff such as product development technologists, packaging technologists, innovation managers and laboratory support staff (i.e. lab technicians). Marketing and research capabilities were also taken into account. The organisations were then slotted into categories (i.e. none, low, medium or high) depending on the level at which they were judged to be.

Table 8.1 Capacity for product innovation[a]

Industry/nature of organisation	Company	In-house technical capabilities for product innovation/new product development (resources and expertise)			
		High	Medium	Low	None
Grocery retail chains	QR	–	–	✓	–
	ST	–	✓	–	–
Manufacturers (milk category)	W3	✓	–	–	–
	W9	–	–	✓	–
Manufacturers (flour category)	W1	✓	–	–	–
	W3	✓	–	–	–
	W12	–	✓	–	–
Manufacturers (cheese category)	W3	✓	–	–	–
	W10	–	–	✓	–
Manufacturers (breakfast cereals category)	W1	✓	–	–	–
	W2	–	✓	–	–
	W4	✓	–	–	–
Manufacturers (tomato sauce category)	W6	✓	–	–	–
	W11	–	–	✓	–

Source: Table compiled for this book based on interview data and other data from documentation

[a]Reprinted from Australasian Marketing Journal, Vol. 23, No. 1, Chimhundu et al., Manufacturer and retailer brands: Is strategic coexistence the norm? Page 55, Copyright (2015), with permission from Elsevier

The column marked "high" in Table 8.1 consists of organisations that have full capacity to carry out their own innovation activities, and as we move towards the low side, the organisations here may have limited capacity for innovation, and therefore either do not do much innovation themselves or rely on other organisations to do it for them, so they are largely followers. The table summarises capacity for innovation for the grocery retail chains and the manufacturers. It should be noted, however, that the columns ranging from "none" to "high" should be seen more as a continuum, since the rigid boundaries were created for the convenience of categorisation. Therefore, even organisations in the same column (say, medium) may still be at different levels on the continuum of capacity for innovation.

It is also important to mention that, on the retail chain side, such categorisation may be more of a reflection of strategic choice and what the chain defines its business mission to be than anything else. For instance, a retail chain's strategic choice not to invest much in R&D facilities and related human resource expertise may be based on the decision to largely leave innovation activities to the manufacturers, a stance that was largely exhibited by retail chain QR. On the other hand, another retail chain may, as part of its mission, want to be involved in more activities related to innovation and new product development, as indicated by a senior manager in retail chain ST: "I guess we are now at a phase in our business where we are quite prepared to take some of the innovation ourselves [...]. It's a domain that we are certainly very interested in" (Interview ST1).

As can be seen from the table, the "high" column is dominated by manufacturers, and all FMCG/supermarket categories in question are represented in this column. However, neither of the retail chains is represented in this column. One retail chain is in the "low" column and the other one is in the "medium" column. The "high" column on the manufacturer side is dominated by large companies. Small to medium companies are generally in either the "low" or "medium" column. What is shown by this table is that manufacturer brands have greater collective capacity for innovation than private label. In addition, nine research participants either mentioned directly or indirectly the superior capacity for innovation on the part of manufacturers.

The retail chain that is in the "low" column has described how use is made of manufacturers' R&D facilities after identifying new product opportunities: "We then go out to our suppliers and they use their resources, their labs, their technical people" (Interview QR2), and the retail chain eventually works with the chosen manufacturer. This demonstrates a strategy and business model that does not overcommit the organisation in areas that are outside the core business functions of retailing, wholesaling and distribution. The retail chain that is in the "medium" column is described as having "a full product development team [...]; one big laboratory [...] as if we were a manufacturer" (Interview ST1). There is a description of its "own scientists and [...] own laboratories" (Interview ST1) at the parent company outside New Zealand; however, there is an admission that from an R&D perspective "Supermarket companies as a whole don't tend to spend a lot of money on R&D. That's more of a branded company [function]" (Interview S3). A comparison of the two retail chains shows that they may have different philosophies concerning their desired level of capacity for innovation. In addition, retailers are described as being largely dependent on suppliers "because they [suppliers] have the technology and the knowledge" (Interview Y1).

Greater collective capacity for innovation would naturally be expected to enable manufacturer brands to innovate at a higher rate than retailer brands. This takes the discussion into the second complementary point of research issue 3a, on the rate of innovation. This is largely assessed through the analysis of interview data. Table 8.2 presents interview data on rate of innovation.

Table 8.2 Manufacturer brand and private label rate of innovation

Views expressed by participants	Retail chain QR (*n* = 21)	Retail chain ST (*n* = 9)	Manufacturers (*n* = 13)	Consultants (*n* = 3)
Manufacturer brands innovate at a higher rate than private label	14	5	8	2
Private label innovation is increasing	–	–	2	–
Private labels innovate at a higher rate	–	–	–	–

Source: Table prepared for this book from research interview data

It is important to note that the nature of this table is such that the comments given on the issues at hand do not have to add up to the number of interviews. For instance, from the 21 interviews in retail chain QR, 14 interview participants gave comments on the issue of manufacturer brands innovating at a higher rate than private label brands. The comments do not have to add up to 21 because the other interviewees may not have said anything about the issue or may have said something that does not really address the issue in a way that enables meaningful reporting. The cross-section of interview participants within the regional company of the retail chain was also such that not all interview participants were in a position to give comments on all issues. For instance, some managers at lower levels would be better placed to give detailed information about certain aspects of products in their categories but would consider themselves not well placed to give comments about issues related to product innovation. Therefore, the reporting of research interview data sought to establish the big picture with regard to issues, and was not structured in such a way that every interview participant would have to give an answer on a particular issue so that a reconciliation of the answers in terms of numbers could be made. This is not seen as detracting from the conclusions that can be drawn from the tables.

The table shows that in the categories, manufacturer brands are innovating at a higher rate than private label. There is overwhelming agreement on this right across the spectrum of different types of research participants: retailer participants, manufacturer participants and consultants. And from manufacturer participants, the eight responses shown in the table straddle all categories studied, thereby providing triangulation across all the categories.

Quotations extracted from interviews are further presented and discussed next, from the viewpoint of who does more innovation in the categories: manufacturer brands or private label. The text evidence in Tables 8.3 and 8.4 crosses retail chains, and crosses all four supermarket groups and all five main categories studied. Therefore, triangulation across all these units, getting more or less similar comments on manufacturer brands innovating at a higher rate in the categories, puts one in a position to be able to claim the robustness of this conclusion, using replication logic. What is interesting to note also is that even within the retail chain that was

Table 8.3 Retailers, text evidence on rate of innovation

Abridged text evidence on manufacturer brands innovating at a higher rate than private label
Retail Chain QR
"They [manufacturer brands] do the work [...] They spent the money [...] The innovation [...] and most of the TV advertising to bring customers into the store is largely driven by manufacturer brands; and then when they get to the fixture, they have a choice of going private label or manufacturer brand" (Interview QR1).
"Probably the proprietary brand [manufacturer brand, innovates more] [...]; and that's their market too. If they don't constantly have new products coming through, dropping off the old, slow ones and bringing in new ones all the time, then they get left behind" (Interview Q1).
"I would say [...] 10 to 20% [is contributed by private label] [...]; negligible. And not in all the categories because there [are some categories] where house brand has invested into and come up with new products to the category" and have been successful (Interview Q3).
"I don't think there is that much work on private label. So it will be like maybe 60:40." (Interview Q4).
"I think you usually find that the ones that are innovating are [...] the big companies anyway [...] but a lot of the time, what happens is because of competition" (Interview Q5a).
"It would be manufacturer brands [that innovate more]." And on the ratio of innovation output of manufacturer brands/private label respectively: "I would say 10:1, maybe more, 15:1" (Interview R1a).
Estimated ratio of innovation in the categories put at "95:5"in favour of manufacturer brands (Interview R2a).
"Probably about 4 to 5% in the house brand [private label]"; manufacturer brands putting in the bulk (Interview R4).
"Most of the innovation and marketing would be done by the proprietary companies" (Interview R5a).
"I would say [innovation is] at least 80 to 90% manufacturer brands" (Interview R6).
Retail Chain ST
There is innovation in all categories, including commodity ones: "Absolutely; even in categories like flour, their innovation could be packaging changes" or some other changes. "As an average, I could probably say that we launch 200 new products every week across our total offer. This is not just private label [...]. So, if you multiply that by 52 that would be your annual number". This covers the whole range of innovation from incremental to radical. On the proportion of innovations between manufacturer brands and private label, the majority would be manufacturer brand innovations: "Private label would be 5%" (Interview ST1).
"They have just brought a few new ones [products] in the [private label]". However, there is a higher rate of such activity on the manufacturer brand side (Interview S1a).
"It's minimal in private label" (Interview S2a).
"The other companies [manufacturers]" do more innovation in the categories than private label (Interview T1).

Source: Table prepared for this book from research interview data

Table 8.4 Manufacturers, text evidence on rate of innovation

Abridged text evidence on manufacturer brands innovating at a higher rate than private label
Milk and Cheese Categories
"There is more innovation from the manufacturer, from the branded product, obviously, because non-branded is actually increasing in its share, and through its price especially in cheese and milk." (Interview W3a).
"Manufacturer brands are well ahead [...]. [The] only new ideas you see in that segment [are from] manufacturer brands" (Interview W10).
Flour and Breakfast Cereals Categories
"Private label brands would not really invest to any great extent at all in the category [...]. So they would be seen as predominantly an everyday low price [option]. There would be nothing outside of pricing, other than, I suppose, the higher-level investment. So [retail chains'] investment in their private label brand occurs more on a level of the [private label] brand rather than specifically [private label] flour" (Interview W3b).
"There has been a lot [of innovation], especially in the cereal category [...]. I spent three years in the UK before coming back to New Zealand, [and UK] private label is very strong. And so you can really see the New Zealand supermarkets are tending to go towards that frame of mind with their products. Yeah, product innovation across both types of brands is always on the go; it's always happening. So, there has been a lot recently in the category [...]. The brands are getting a lot cleverer in their packaging, which is really helping their brand. Before, we used to have quite basic private label brands, but now both [names of private label brands] and [the respective supermarket chains] have really done a great job on the packaging; so is [name of different private label brand] with the likes of [name of retail chain] as well" (Interview W1).
"The private brand is always going to be the one to deliver the innovation because they have to sell the product at a higher price than the house brand" (Interview W12).
"Manufacturer brands drive the category, bring innovation and generally drive promotion. So it's not the role of store brands. Store brands are a follower". Asked for a rough estimate of relative dollar input into innovation by private label brands compared to manufacturer brands, the same participant said: "Nothing; private labels don't innovate". On product innovation and promotion: "There is [...] almost no innovation in private label" (Interview W2).
"The house brands generally don't have innovation. They are normally followers of trends, and when a trend becomes successful, they tend to create a home brand based on the success of that trend." The innovation is therefore largely done by manufacturer brands in the category, according to this participant. On the relative contribution of manufacturer brands and private label to category investment: "I don't know the percentage but the home brand [private label] would be a small percentage" (Interview W5).
Tomato Sauce Category
"In my opinion, it's all driven by the brands [manufacturer brands] at this stage [...]. It's certainly [the case] from a market research type point of view, but product development type [...], as well as general market research, generally in our category is very much driven solely by the brands" (Interview W6).

Source: Table prepared for this book from research interview data

described in an earlier section as having a more aggressive and ambitious stance on innovation, the private label rate of innovation in proportion to that of manufacturer brands was described by a key manager as constituting only a small percentage: "Private label would be 5%" (Interview ST1).

Most of those who provided an estimate percentage of private label innovation gave figures that are not far from this, thus providing the necessary triangulation. All this supports the proposition that there is a greater collective capacity for innovation on the part of manufacturer brands, resulting in a higher rate of innovation by them in the categories. Only one of the research participants made comments about private label making strides in innovation, and this was from the viewpoint of product and packaging quality improvements, as private label brands have traditionally been seen to be of low quality: "The brands are getting a lot cleverer in their packaging, which is really helping their brand. Before, we used to have quite basic private label brands" (Interview W1). So, private label brands are improving in this area.

The results in this section on capacity for innovation and rate of innovation have shown that the resources and expertise of manufacturers enable them to collectively contribute more towards innovation in the categories than private label brands do (Chimhundu et al. 2015). Regarding research issue 3a (How do private label and manufacturer brands compare on capacity to innovate in the FMCG product categories?), the discussion in this section gives rise to the following research proposition:

Research Proposition 3a: Due to their resources, manufacturer brands have superior collective capacity to innovate at a higher rate than private label.

The term "resources" is meant to incorporate not only facilities, finance and the like, but also expertise (i.e. expert resources).

Analysis of Research Issue 3b

Research Issue 3b: How do private label and manufacturer brands compare on capacity to contribute to category marketing support and development?

Table 8.5 Manufacturer brand and private label contributions to category development/category support

Views expressed by participants	Retail chain QR ($n = 21$)	Retail chain ST ($n = 9$)	Manufacturers ($n = 13$)	Consultants ($n = 3$)
Manufacturer brands make a greater contribution to category development (manufacturer brand contributes the bulk)	10	3	7	–
Both manufacturer brand and private label make a contribution	1	–	2	–
Dollar cost of promotion built into pricing/cost structure for manufacturer brand, unlike private label	3	–	2	–
No need to promote private label/depends on the category	3	3	–	–
Private label makes a greater contribution to category development	–	–	–	–

Source: Table prepared for this book from research interview data

Interview data on capacity for category development are presented in Table 8.5. Although product innovation does develop the categories, such innovation was analysed separately as it is an important facet of this research. Category development in this section of research issue 3b involves looking at all marketing aspects other than innovation that help to develop and grow the categories. A complementary term that is used in this book for such category development activities is category marketing support.

It is shown in this table that manufacturer brands contribute the bulk of the category support that develops and grows the categories. This aspect received a comparatively high number of comments, ranging across grocery retail chain QR and its respective supermarket groups, grocery retail chain ST and its supermarket groups, and manufacturers as well. On the retailer side, the comments naturally range across all categories because the retailers studied deal in all the categories covered by the research. On the manufacturer side, the seven comments in support of manufacturer brands making a greater contribution to category support than private label range across all the relevant categories; milk, flour, cheese, breakfast cereals and tomato sauce. In addition, none of the research participants talked to were of the alternative view that private label could be making a greater contribution to category support than manufacturer brands, which seems to indicate that this aspect is well known in the FMCG industry.

In terms of specific comments, the following two are representative of the majority of comments on the issue of manufacturer brands making a greater contribution to category development. They are derived from both the retailer and the manufacturer camps.

> Most of the TV advertising to bring customers into the store is largely driven by manufacturer brands; and then when they get to the fixture, they have a choice of going private label or manufacturer brand. (Interview QR1)

> The private label marketing activity is extremely limited […]. They undertake very little above-the-line advertising. So, my estimate would be, the marketing activity […] by private label is very low, on the whole. (Interview W4)

A large component of category development expenditure is linked to advertising and private label is not really a big player in this area. Advertising develops awareness, educates consumers about brands and products and persuades them to buy, but it is a costly activity. From the point of view of spend and share-of-voice, manufacturer brands are regarded as big contributors to category development. This book takes the view that direct brand support is also category development/category support in the sense that brands make categories and the success of brands leads to the growth of categories. Therefore, any activities geared towards researching and understanding market segments or customers in a particular category, or promoting new or existing brands, are integral to category development.

On the matter of expenditure on category development, there is also the view that private label has to be competitive on price, and that it would not be able to charge competitive prices if it engaged in heavy advertising and promotional activities, which would have to be funded from somewhere. The consumer would therefore have to bear the cost and that would defeat one of the primary purposes of the private label; that of providing affordable alternative products. Naturally, therefore, private label strategy is such that it cannot afford to do both.

> Now, the nature of private label means that we have to be more competitive in our price offering. As a result, we don't have the same level of spend for doing above-the-line activities like TV advertising, magazine advertising and those sorts of things. (Interview QR2)

Furthermore, there is a general conviction that manufacturer brands do have the resources to engage in above-the-line activities and a lot of the category development activities because the costs associated with such expenditure are already built into the cost structure/price structure of their products, unlike private label. Therefore, manufacturer brands should be able to afford more for such types of activities:

> Proprietary brands have that [dollar cost] built in, so they can do their advertising budgets and do all their advertising; and what we do hope is that, that would overall grow the category and allow us to maintain a feasible position in there as well. (Interview QR2)

Nevertheless, some changes have been noted in the New Zealand FMCG industry relating to private label category support activities, as indicated in this excerpt: "Advertising and sales promotions: we are seeing more done by private label than we did in previous years" (Interview W6). This is merely a change that reflects more promotional activities by the private label from a previous state of affairs in which the private label was hardly advertised or promoted. This, however, does not in any way imply that the private label is catching up with the manufacturer brand on category support because the proportion of private label contributions to category support is still considered to be small: "In my category [...] I would say they [private label brands] probably do 10%" (Interview W6).

Triangulation with the estimates supplied by other research participants who commented on the issue confirms the disproportionate balance from a category development perspective. Estimates of manufacturer brand/ private label comparative contributions to category development are put at: “I would […] say 10 to 20% […]; negligible” (Interview Q3); “80% towards manufacturer brand really” (Interview R1a); “More than 90%” is manufacturer brand contribution (Interview ST1). Interview W2 puts the estimate of manufacturer brand contributions at “90%” or even higher. What is interesting to note is that even in retail chain ST, which was discussed earlier as seeming to have more ambitious private label plans, private label contribution towards category support is still put at a low 10% (Interview ST1 above).

In addition, it should be noted that the private label stance towards category support activities may vary from category to category. For instance, while there is promotional activity in some categories, there is not in others, as evidenced by these sample quotations from the two different retail chains.

> We do [promote] cheese; we don’t promote milk because there is not a lot of money in it. If you start fighting with milk, milk and bananas are a massive part of your shop; everyone buys milk. So if you go and promote milk, it means that [the opposition retail chain] are going to promote it, and every week if they promote milk we have to promote it; so you take a lot of money out of it. So, you will never make money if you start promoting products like that. (Interview R4)

> Normally milk doesn’t really get put on special all that often. It’s generally the price that it is… in our weekly catalogue… sometimes we… put milk in there but not a lot because milk is something that people come to buy. It’s not an impulse buy. (Interview T3)

It seems that whatever promotional activities go on in such categories, they are carried out largely by manufacturer brands. Even in categories where the private label is promoted, it has already been established that manufacturer brands do the bulk of the category development activities.

Category support activities require a lot of financial resources and expert support, and the collective resources of the different suppliers ploughed back into the categories through category support are greater in magnitude than those of private label. Therefore, with respect to research question 3b (How do private label and manufacturer brands compare on capacity to contribute to category marketing support and development?), this discussion gives rise to the following research proposition:

Research Proposition 3b: Due to their resources, manufacturer brands have superior collective capacity to make a greater contribution to category marketing support and development than private label.

Stance on Comparative Capacity for Product Innovation and Category Support

Analysis of Research Issues 4a and 4b

These two research issues are tackled simultaneously in this section.

Research Issue 4a: What is the state of awareness of FMCG retailers on the comparative capacity of private label and manufacturer brands to innovate and give marketing support to the product categories?

Research Issue 4b: What is the strategic stance of the FMCG retailers with respect to this comparative capacity to innovate and give marketing support to the product categories?

Table 8.6 shows data relating to awareness in the FMCG industry, of the status of capacity for innovation and category support on the part of manufacturer brands in relation to private label.

As shown by the numbers, it is clear that there is a general awareness within the industry that the capacity of manufacturer brands to drive innovation and development/support within the categories is superior to that of private label. This view is largely shared in both retail chains QR

Table 8.6 FMCG retailer awareness and stance on manufacturer brand superior capacity for innovation and category support

Views expressed by participants	Retail chain QR (*n* = 21)	Retail chain ST (*n* = 9)	Manufacturers (*n* = 13)	Consultants (*n* = 3)
Aware of manufacturer brand superior capacity for innovation and/or category development	16	7	10	2
Private label expects positive gains from that capacity	14	4	10	2
There are perceived benefits for manufacturer brands in the process (e.g. capacity utilisation)	4	–	3	–

Source: Table prepared for this book from research interview data

and ST, and within the different operational units of the grocery retail chains that include the supermarket groups selected for the study and head office operations. The view is also shared by manufacturers in all the categories covered, as well as by consultants. This awareness is consistent with factors established earlier; factors that include the dominance of manufacturer brands in most categories, the differences in resources and expertise for innovation and category support between manufacturer brands and private label, and the differences in rate of innovation between the two types of brands.

It should also be noted that most comments relating to previous tables relevant to research issues 3a and 3b about manufacturer brands having the collective capacity to innovate at a higher rate than private label and manufacturer brands having the collective capacity to make a greater contribution to category development than private label were in fact made by the retailers themselves.

Moving on from awareness, Table 8.6 further reports on how retailers perceive the private label's coexistence with the manufacturer brand with regard to the disparities on contributions to category innovation and category support. While the two are in competition, there is an element of dependence between them. A lot of the comments were given about retailers either directly or indirectly having a stake in manufacturer brands' superior capacity for innovation and category development, as the benefits of such activities flow through to the private label as well. Typical comments given within retail chain QR on this issue include:

> Bear in mind that private label is traditionally not innovative in that we don't enter into categories that are not already developed; so, innovation for us comes down to looking for new product ideas, mainly within existing categories. (Interview QR2)

> You need that branded product [manufacturer brand] to drive innovation. You need that branded product to drive promotional programmes. You need the branded product to generate advertising revenue and cooperative dollars [...]; all those sorts of things. (Interview QR3)

> The manufacturers' brands have a lot more money invested in probably carrying the other house brands, but you couldn't have it the other way round. (Interview R2b)

Entering into categories that are "developed already" by manufacturer brands; relying on manufacturer brands to "drive innovation [...]; drive promotional programmes"; and the money invested by manufacturer brands that enables the "carrying [of] house brands" are all indications of private label drawing upon the capacity of manufacturer brands in the areas of innovation and support in the product categories. A lot of similar comments were made regarding this aspect. While there may not be any explicitly articulated intentions (either verbally or in writing) on the part of the retailers and the manufacturers on this aspect, the mere pattern of activities supports a situation where retailers are seeking to draw upon manufacturer brands' superior capacity for innovation and category support to drive the categories. Driving the categories benefits both private label brands and manufacturer brands.

Within the other retail chain studied, retail chain ST, a similar pattern was expressed indirectly by a senior manager: "Historically, suppliers have owned the domain of innovation, and private label has probably followed" (Interview ST1). This aspect of following on the footsteps of brands that bring about pioneer offerings and activities is consistent with drawing upon what has been put in place in the categories. Typically, manufacturer and consultant comments concur on this issue. The following comment was made by a manufacturer:

> I think what happens there is, once the manufacturer grows the category, and that could be through innovation, then some of the house brands would pick up on that innovation and go into that particular product to assist their overall growth in the category and maintain their share of that category. (Interview W5)

The following comment was made by a consultant: "Private label have [...] a tendency in general to pick up the ones that are successful" (Interview Y1).

Nevertheless, this situation cannot be seen as one-sided, as manufacturer brands also derive benefits from this relationship with private label. Across the interview participant spectrum, nine comments were made relating to manufacturer brands indirectly benefiting from the process; for instance, as spelt out here:

> But then you are talking about a double-edged sword as well. What happens is, they [manufacturers brands] innovate, they grow the category and then house brand comes along [...]. So, they want to get on board with the house brand, [and say] if it's going to be a house brand, we will produce it. (Interview Q3)

> The biggest thing that we offer manufacturers is volume, which helps them with their viability in terms of running their production lines because they have additional volume offered through private label which they wouldn't have if they weren't doing private label. (Interview QR2)

With aspects of this nature coming into the equation, therefore, the situation shifts from a general outlook of private label brands that are overdependent on the innovation and category support activities of manufacturer brands, to ones that are depended on by manufacturer brands as well. It is not just a situation of private label drawing upon the capacity, but rather of it doing so as part of a strategic interdependence. This is tantamount to both types of brands contributing to the pool in one way or another and both drawing from the pool. This section has addressed research issues 4a (What is the state of awareness of FMCG retailers on the comparative capacity of private label and manufacturer brands to innovate and give marketing support to the product categories?) and 4b (What is the strategic stance of the FMCG retailers with respect to this comparative capacity to innovate and give marketing support to the product categories?).

From the discussion in this section, therefore, two research propositions arise, respectively:

Research Proposition 4a: Retailers are aware of manufacturer brands' superior capacity for innovation and category development.

Research Proposition 4b: Retailers seek to draw upon this capacity as part of a coherent, pooled interdependence strategy.

A further discussion on how good or bad manufacturer brand activities are for private label brands becomes necessary in order to perform a more comprehensive analysis of the strategic dependency aspect.

The fact that manufacturer brand and private label products are sitting side by side on grocery retail shelves, competing for customers, would tend to give the impression that any competitive activity on the part of one brand hurts the other. So, one might reasonably assume that the innovation and category support activities of manufacturer brands always hurt private label. A systematic analysis of research interview data for relevant themes in the area of product innovation (and marketing support) has shed more light on the impact of manufacturer brand innovation on private label. Is the impact largely negative or positive for private label? In other words, does manufacturer brand innovation enhance or inhibit private label in FMCG/supermarket product categories? And are there aspects that are consistent with "drawing upon" the other category participant's capacity as discussed above? The research unveiled one theme related to the inhibition of private label by manufacturer brands and three themes related to the enhancement of the private label by manufacturer brands. These themes are discussed below, with supporting representative quotations from the research interview data.

The first theme is represented by the following set of similar quotations:

> You can get innovations by proprietary brands that we don't have access to in private label, and then that can inhibit the growth of private label because people tend to head towards the innovation and what's new. (Interview QR2)

> There is more innovation from the manufacturer: from the branded product, obviously, because non-branded is actually increasing in its share, and through its price, especially in cheese and milk. (Interview W3a)

> More private label penetration would see more innovation happening [...], trying to reverse the trend of private labels. (Interview W10)

> Private label is seen as trailing "the supplier who has to constantly innovate to stay ahead of the private brand". (Interview Y1)

The theme that can be derived from these quotations is: *The competitiveness of innovative manufacturer brands inhibits private label* (Theme 1). Manufacturer brand innovation is perceived as posing a competitive threat to private label in the categories, either by way of bringing in products that are superior and more appealing to consumers, or by way of staying ahead of private label on innovative activities, or through intensifying innovative activities in order to reverse the gains of the private label. In this way therefore, manufacturer brand innovation can be said to have a negative impact on private label.

The second theme relates to the following quotations:

> Manufacturer brands drive the category, bring innovation and generally drive promotion. So, it's not the role of store brands. Store brands are a follower. (Interview W2)

> If you leave that to private label, there is no funding for it; so you are not going to drive the category. Anybody who is going ahead in manufacturer

categories, you are generally spending money on that product somewhere, whether it would be in advertising or innovation of new products or in-store dealing [...]. They [manufacturers/manufacturer brands] are trying to grow their category as a whole, and by that, we [retailers/retailer brands] just get sucked along in the vacuum. (Interview R2a)

But then it [manufacturer brand innovation] can also enhance private label because we can go into categories where there has been innovation and we can come along with private label and offer the same at a more competitive price [...]. (Interview QR2)

[Manufacturing companies are] investing in the categories and new products ... if you go down and look at the sauces category now, you will see sauces that weren't there years and years ago [...]. Barbeque, sweet and sour; all the investments that these companies have been putting into the brands to expand the sauces range. And while it has diluted the sales of the standard tomato sauce, it's growing the category. (Interview Q3)

The theme that can be derived from the above quotations is: *Private label brands are enhanced by manufacturer brand innovation driving the categories and category growth* (Theme 2). Manufacturer brand innovation is perceived as driving the growth of grocery retail categories and thereby benefiting not only the manufacturer brand but the private label as well. In this regard, there is an element of the private label being carried by the manufacturer brand. In addition, manufacturer brand category investment that benefits private label is not limited to product/brand innovation, but extends to the marketing activities associated with the innovations as well.

The third theme is represented by the following set of comments:

"Most house brand products are copies of our mainstream branded products. There [are] a couple that aren't, but generally that's what they are, and it probably relies on the big brands to do the work" (Interview Q3). "The house brands generally don't have innovation. They are normally followers of trends, and when a trend becomes successful, they tend to create a home

> brand [house brand] based on the success of that trend" (Interview W5). "The market leaders [manufacturer brands] […] do the innovation. Private label follows". (Interview Q6)

> Bear in mind that private label is traditionally not innovative in that we don't enter into categories that are not already developed; so, innovation for us comes down to looking for new product ideas mainly within existing categories. (Interview QR2)

> Innovation for us comes down to [identifying] new product opportunities […]. We then go out to our suppliers and they use their resources, their labs, their technical people to come up with a proposal. Then they come back to us and they submit it, and we may have two or three suppliers who are doing that. And then we select the one that we want to work with. (Interview QR2)

> Typically, what they [retailers] would do is approach a supplier to come up with formulation for them […] or they might get a favoured supplier to submit some samples and then say, okay, we want this made according to this recipe […] because they [suppliers] have the technology and the knowledge. (Interview Y1)

The theme that can be derived from the above quotations is: *The adaptation of successful manufacturer brand innovations by private label enhances private label* (Theme 3). Manufacturer brands lead on innovation and private label largely follows by way of adapting/copying successful manufacturer brand/product innovations, or through entering into categories that have been developed already by manufacturer brands. That way, the cost and risk of private label "innovations" is reduced. Furthermore, it is the manufacturer brand resources such as labs, technology, specialised staff and knowledge that play a big part in bringing about the innovations, and in many cases adapting them or supplying them for private label requirements should the private label wish to introduce its own of the same nature.

The fourth theme comes out of the following set of representative quotations:

> If you are innovating within products, there is generally a bit of money in new products, so you can make a bit of margin out of it, and it can bring you a bit of interest in a category which can be stagnant. (Interview Q3)

> They [manufacturers] put the money/spend into it. As house brand, we ride off the back of it, hoping that the manufacturers drive people into the category and then they buy our house brand while they are there" (Interview R4). "It [manufacturer brand innovation and marketing] drives consumers to the category" (Interview QR1). "The brands bring people to the store; private label doesn't do that. (Interview QR1)

The theme that can be derived from these quotations is: *The customer pulling power of innovative manufacturer brands enhances private label* (Theme 4). The interest and customer pull generated in a category by leading and/or innovative manufacturer brands benefits both the manufacturer brand and private label as private label brands are displayed alongside manufacturer brands.

This discussion therefore demonstrates that although the competitiveness of innovative manufacturer brands inhibits retailer brands, manufacturer brand innovation and category support actually have a huge positive impact on the welfare of private label brands on aspects related to driving category growth, private label adapting successful manufacturer brand innovations, and private label tapping into the customer pulling power of innovative manufacturer brands (Chimhundu et al. 2010, 2015). These three aspects that have a positive impact on the welfare of private label brands in the categories give rise to proposition 4b about retailers seeking to draw upon manufacturer brand capacity for innovation and category support as part of strategic dependency.

In addition, it is therefore reasonable to argue that strategic management regimes governing the coexistence of the two types of brands in the categories are partly influenced by retailer perceptions of manufacturer brands' ability to positively impact on private label. This leads the discussion into the next section, which deals with research issue 4c.

Analysis of Research Issue 4c

Research Issue 4c: How is this strategic stance related to policies on the coexistence of manufacturer brands and private label in FMCG product categories?

The perceived implications of a possible overdominance by private label in FMCG/supermarket categories as per the research interview data are shown in Table 8.7.

Table 8.7 Implications of private label overdominance (product innovation and category support perspectives)[a]

Views expressed by participants	Retail chain QR (*n* = 21)	Retail chain ST (*n* = 9)	Manufacturers (*n* = 13)	Consultants (*n* = 3)
Overdominance of private label would have a negative impact on category innovation and/or support	10	3	4	2
Private label share rise would mean more innovation and category support to reverse the trend	1	–	3	–
Innovation in some categories not entirely a function of New Zealand based innovation	–	1	–	1
Overdominance of private label brands would have a positive impact on category innovation and/or support	–	–	–	–

Source: Table prepared for this book from research interview data

[a]Reprinted from Australasian Marketing Journal, Vol. 23, No. 1, Chimhundu et al., Manufacturer and retailer brands: Is strategic coexistence the norm? Page 56, Copyright (2015), with permission from Elsevier

Spelt out in the table is strong support for the view that an overdominance of private label in the categories would have a negative impact on category innovation and/or category support on the part of manufacturer brands. Effectively, this means a negative impact on overall category innovation and/or support since, as established already, manufacturer brands contribute the bulk of it. Similar comments range across the two retail chains and the four supermarket groups studied, as well as across four categories on the manufacturer side and two consultants. Typical comments include:

> If private label becomes too big, then the proprietary brand will withdraw from marketing, innovation [...]; and that's where it all comes from because private labels aren't innovators. (Interview QR1)

> If private label is too dominant, you don't get the same degree of new product development and the onus then ends up on the retailer to develop the new products. And then you get the whole lot of cost structures coming into the business [...]. (Interview QR3)

> If it [private label] becomes too dominant, then it slows down the growth of the category [...]. They [manufacturers] are going to spend money to make their product number one or they are going to spend money to invest in new technology or new products or something like that, which keeps the category alive; [...] our house brand comes along behind and just gets dragged along, driven up by their spend at no cost to us. (Interview R2a)

> At the end of the day, as we head more and more into private labels, the amount of innovation and development that's coming from the manufacturers will decline because the manufacturers then don't have the money to invest in research and development. (Interview W3b)

> That will kill innovation long term [...] because the profitability won't be there for the supplier. So, they have got to watch out that they don't kill their source of innovation; otherwise the whole thing grinds to a halt. (Interview Y1)

The perceived negative impact on innovation and marketing support on the part of manufacturer brands due to diminished resources and lack of incentive to plough back into the categories could eventually hurt the categories, along with the category participants, retailers included. It therefore seems that there is no long-term strategic advantage for the retailers of private label overdominance of the categories. One might reasonably argue that the contributions of manufacturer brands towards innovation and marketing in the categories do play an important role in the determination of the level to which the private label should grow, and in the balance between manufacturer brands and private label in the categories. This issue hasn't really been articulated in the mainstream academic literature, especially in comparison to a factor such as consumer choice which will be covered in another section.

From a retail chain policy point of view, manufacturer brands form an integral part of private label marketing strategy because the private label has open lines of strategic dependency with the manufacturer brand. This gives further support to the existence of long-term equilibrium points in the categories, beyond which manufacturers may not want to go with respect to private label growth and share.

As has been noted from in-store category observation data and category share data, private label penetration is much higher in commoditised categories such as milk and flour than in non-commoditised ones. However, it is not envisaged that the situation of a 100% private label share is desirable to the retailers, even in such categories. There is still an optimum level of penetration that is desirable, beyond which the long-term performance and growth of the category comes under threat. Even in such commoditised categories, the argument linked to the threat on manufacturer brand category innovation and support is still applicable, as shown by the statement from a senior manager in the retail chain that was reported as being ambitious for its private label that "[even in] highly commoditised categories, it's not our intent to increase our share to over the dominant position" (Interview ST1). Despite the fact that this manager notes the increased role of private label in the areas of innovation and marketing support, the element that is contributed by manufacturer brands is seen as critical to the prosperity of any category. For instance, the manager expressed this view:

> Historically, suppliers have owned the domain of innovation, and private label has probably followed [...]. I guess we are at a phase now in our business

> where we are quite prepared to take some of the innovation ourselves […]. It's a domain that we are certainly very interested in. (Interview ST1)

The senior manager added that the approach would be "Probably more selective in categories". But despite the comments above, the manager still saw the role of manufacturer brand innovation and marketing in such categories as critical, and dismissed the possibility of a 100% private label brand share:

> No […]. Because there is still a role for innovation and technology in terms of milk. You can look at all the [brand name of milk] […], all the […] added-value parts of milk. I guess we need to continue to have total category innovation so that they [manufacturer brands] supply […] private label. (Interview ST1)

On the same issue, another manager in the retail chain talked about the idea of taking care of suppliers: "You still have got to look after the suppliers because at the end of the day, they have got a place in the market as well, and it's to no one's benefit to get rid of the suppliers" (Interview S3). Yet the interviewee also made comments to the effect that if the retail chain really wanted to, it would be possible to do without supplier brands in certain categories. For instance, in milk, "We could do away with the main suppliers. We don't really have to have them. All we would have [would] be the bits and pieces like the [names of brands] and stuff"; and in flour; "You could in flour, if we really wanted to […] we could delete branded flours and have one branded flour" (Interview S3). This shows that the retail chains would do away with suppliers in some categories if they wanted to, but they are not willing to do so. Therefore, against this background of looking after suppliers because they make contributions to the categories, the role of manufacturer brands is seen as important, especially their role in bringing about product innovation and category marketing support, as well as in supplying private label.

There is also the view that high private label penetration would spur manufacturer brands to innovate more in order to outcompete the private label. One comment to this effect was: "More private label penetration would see more innovation happening […], trying to reverse the trend of private labels" (Interview W10). In this way, innovation would be used as a competitive tool. While this can happen, what tends to weaken such a line of thought is that the retailers are the owners of the shelves and they do have the final say. They can delist an innovative brand if they want to, in the process of rationalising the categories. The bottom line, therefore, is that retailer strategic thinking with regard to the contributions of manufacturer brands to category innovation and support has a key part to play in the coexistence of the two types of brands in FMCG categories.

An additional dimension to this discussion is a point mentioned by two research participants on the issue of private label overdominance and impact on category innovation and support. This is reflected in the comments below:

> In the New Zealand context [...] if you sat down and said, who are the top 20, top 30 suppliers to New Zealand supermarkets, most of them are foreign owned [...]. Their innovation in New Zealand or even in Australia is not really a function of New Zealand or Australia. It is a function of some kind of global countries of excellence where they try and continuously develop new products. (Interview Y3)

> When you have international ownership of the multinationals like Kellogg's, they still have that power to innovate at head office level and push it all back down through each company; so, it doesn't change that [...] they still have this enormous ability to leverage the cost of that development. Whereas the small companies, we develop something and we are dependent on purely the New Zealand market size to gain a return on the investment that we have put in; it gets very hard. So, it depends who you are. (Interview W2)

The logic of these two comments is that, an overdominance of New Zealand supermarket categories by private label brands would not have a negative impact on the brand/product innovation of multinational brands, since such innovation is not really driven from within New Zealand but from overseas. In a way, the argument holds water, especially in categories where the majority of manufacturer brands that participate are multinational in nature. But even then, an extremely marginalised multinational brand would have no incentive to continue pouring resources into a category in which it is not reaping any benefits. The situation would then not help the remaining private label. Secondly, there are quite a number of categories that are largely domestically supplied, and whose innovative and category support activities are largely New Zealand based or dependent on the New Zealand market. This calls for a quick analysis of the categories that are the focus of this research to see if the supply and innovation is largely domestically driven or foreign driven. This information is given in Table 8.8. The data were compiled from brands that were observed on the shelves during the category observation exercises.

The milk, flour, cheese, breakfast cereals and tomato sauce categories are largely domestically supplied from within New Zealand. Where a subsidiary of a multinational concern exists in New Zealand, the supply was taken as domestic, although of course it should be noted that from a brand/product innovation perspective, a good part of the work would be coming from outside the country.

The dominant domestic supply situation as evidenced in Table 8.8 backs up the arguments advanced in this section on the negative impact on innovation and category for the five categories of milk, flour, cheese, breakfast cereals and tomato sauce.

This section that analysed research issue 4c and made the argument that strategic management regimes that govern the coexistence of manufacturer brands and private label in grocery retail categories are influenced by the ability of manufacturer brands to enhance private label (Chimhundu et al. 2010, 2015). Manufacturer brand innovation and category support largely have a positive impact on private label brands in grocery retail categories, and private label brands are largely strategically dependent on manufacturer brands in these areas. Therefore, with respect to research issue 4c (How is this strategic stance

Table 8.8 Supply situation for manufacturer brands in the five food categories studied

Product category	Brands	Suppliers	Domestic/overseas supply
Milk	13 brands	10 suppliers	All domestic supply
Flour	10 brands	7 suppliers	9 brands domestic; one overseas supply
Cheese	9 brands	6 suppliers	8 brands domestic supply; one overseas supply
Breakfast cereals	17 brands	14 suppliers	13 brands domestic supply; 4 overseas supply
Tomato sauce	8 brands	5 suppliers	All domestic supply

Source: Table prepared for this book from documentation supplied and category observation data

related to policies on the coexistence of private label and manufacturer brands in FMCG product categories?), the following research proposition is advanced:

Research Proposition 4c: This aspect of strategic dependency has relevance for the determination of policies that govern the coexistence of private label and manufacturer brands in FMCG product categories.

The next section tackles consumer choice.

Role of Consumer Choice

Analysis of Research Issue 5

Research Issue 5: What role is played by consumer choice in shaping the coexistence of private label and manufacturer brands in FMCG product categories?

With both manufacturer brands and private label serving and competing for the same consumer, this research considered the consumer to be central to the relationship between manufacturer brands and private label. It was therefore necessary to assess the importance of consumer choice in the determination of policies governing the coexistence of manufacturer brands and private label in FMCG product categories. The research evidence on the importance of consumer choice is outlined in Table 8.9.

In this regard, overdominance by private label in the categories is generally perceived as having a negative impact on consumer choice, which in turn has a negative impact on category performance. A combined portfolio of manufacturer brands and private label is seen as necessary for there to be adequate consumer choice, so the coexistence of the two types of brands is essential.

The comments below are drawn from the entire research interview participant spectrum, including retail chain QR and its respective supermarket groups, retail chain ST and its respective supermarket groups, manufacturers from various FMCG categories and consultants. The following are examples of typical quotations on the topic of consumer choice:

Table 8.9 Consumer choice and the balance between private label and manufacturer brands

Views expressed by participants	Retail chain QR ($n = 21$)	Retail chain ST ($n = 9$)	Manufacturers ($n = 13$)	Consultants ($n = 3$)
Overdominance of private label would have a negative impact on consumer choice	14	4	4	1
Important to balance what is good for the business and what is good for the consumer	1	2	–	–
Branded products (manufacturer brands) more progressive in addressing consumer wants	–	1	–	1
Close link exists between consumer choice and innovation within the categories	–	–	1	–

Source: Table prepared for this book from research interview data

> Their [retailers'] main focus is providing a range of products for consumers, and the consumer wants selection; and therefore they want various brands to be able to [select] from [...]. The category needs a balance of core products that the consumer wants. So, [a] change in balance could potentially lower the returns the retailer will be able to achieve. (Interview W5)

A balance between manufacturer brands and private label in the categories is seen as necessary:

> "Because it's not just about profit; it's about range and choice" (Interview S2a), and "Consumers want to come into our stores and have choice [...]; they want private label offering and branded players" (Interview QR2).

> You want private label as high as possible if you can. We make good money out of the private label and it's good value as well [...]. The profit is good for us and there is the good side for the customer as well; so the more we can sell it the better, but not to a point where we just have [name of private label brand] only on the shelf; we need the choice. (Interview Q5c)

> There will always be customers that will come in and they want a specific brand of flour and they have their mind set on that. A lot of them won't buy house brand products because they believe they are inferior, which they are not. But if we just had house brand there, the category would go downhill. (Interview R1a)

An overdominance of private label would mean less choice for consumers and this is seen as suboptimal. An optimum category portfolio should have an appropriate combination of both. An overdominance of manufacturer brands would not be appropriate either, as the private label offers reasonable quality at affordable prices, which manufacturer brands do not usually offer. Making the highest level of profit possible is best for the business. Ensuring that there is continuous innovation in the category is good for the consumer and the business, and having a wide range of brands and products in the category is good for the consumer from a choice point of view. The extremes of either placing too much focus on the financial benefits of the business at the expense of the consumer, or too much emphasis on consumer welfare at the expense of category financial returns, are both suboptimal. There has to be a balance and this balance cannot be achieved in a category that is overdominated by private label: "I guess there is a balance between what's best for the business also and what's best for the customers" (Interview R6). In addition, "The brands [manufacturer brands] bring people to the store; private label doesn't do that […]. In many categories, customers are brand loyal" (Interview QR1).

Furthermore, one research participant perceived a link between consumer choice and product innovation. Consumer choice is seen as "kind of related in a way to product innovation […]; we look at what consumers want and innovate accordingly" (Interview W6). In this way, manufacturer brands are seen as having the capability to respond to consumer requirements, a point that additionally makes their participation in the categories worthwhile from a consumer choice point of view.

> The consumer is always [becoming] more sophisticated in what they are wanting and desiring from their products and what they are actually wanting their products to deliver for them […]. Branded products [manufac-

> turer brands] are more progressive in [...] developing and satisfying the consumer's wants and needs just because manufacturers, ourselves, have to expand our products that way. (Interview W3a)

> As a house brand, you don't have that choice; you don't have that speciality option [...]. You won't get specialised milk out of your house brand. (Interview W9)

Despite the benefits that private label offers to both the consumer and the retailer, an overdominance of the private label brand in the category is not seen as desirable, and there is a strong view that consumer choice considerations are a valid consideration in the determination of the balance between manufacturer brands and private label. With respect to research issue 5 (What role is played by consumer choice in shaping the coexistence of private label and manufacturer brands in FMCG product categories?), the discussion in this section gives rise to the following research proposition:

Research Proposition 5: Consumer choice considerations have relevance for the determination of policies that govern the coexistence of private label and manufacturer brands in FMCG product categories.

Nature of Coexistence between Private Label and Manufacturer Brands

Analysis of Research Issue 6

Research Issue 6: What is the nature of the coexistence relationship between private label and manufacturer brands in FMCG product categories and how is it driven?

The areas of strategic policy on the coexistence of manufacturer brands and private label in FMCG/supermarket product categories investigated in this book include: private label growth/share and related equilibrium, driving category growth through product/brand innovation and category marketing support, category management arrangements, shelf-related matters and the private label quality spectrum. These are discussed within each category, and then comparisons are made across the categories. A brief description of each is given next.

The first area of strategic policy is private label growth/share and related equilibrium, which is to do with the level to which the private label should grow in the categories in relation to the manufacturer brand from a market share point of view. This also incorporates the related aspect of equilibrium between the two types of brands as well as possible private label overdominance. The long-term equilibrium point is the optimum level to which the private label should grow, and this point is perceived to safeguard the long-term strategic health of the category. The second area, driving category growth through product/brand innovation and category marketing support is to do with whose responsibility it is to drive the categories with the said activities: whether it is seen as the responsibility of the private label or the manufacturer brand, or both. The third, category management arrangements, deals with the issue of who takes responsibility for the management of the categories and how much say they have in category decisions. The fourth, shelf-related matters, is about merchandising issues relating to shelf space, shelf position, brand/product range and stocking decisions as they relate to manufacturer brands and private label in the categories. The fifth, private label quality spectrum, is to do with the quality/price segments covered by the private label brands in the categories in relation to manufacturer brands.

Regarding research issue 6, the following aspects are worth taking note of. Firstly, private label growth and equilibrium differ depending on the category. Retailer expectations of the optimum balance between manufacturer brands and private label in the categories are not the same for milk, flour, cheese, breakfast cereals and tomato sauce. While known factors such as consumer choice and category characteristics (e.g. commoditised vs. non-commoditised) have something to do with it, strategic dependency in the areas of innovation and category support were found to play a key role as well. Secondly, shelf measures and merchandising decisions concerning manufacturer brands and private label also differ depending on the category. Thirdly, category management arrangements follow a more-or-less similar pattern across the categories, except that the criteria for selection of CCs (Category Captains) may vary and the level of control by the different retail chains may differ. It ultimately rests on how the retailer wants to see the category being run.

Fourthly, the issue of the manufacturer brand driving category growth through innovation and category support is common across the categories. It was established that it is largely the responsibility of the manufacturer brand to bring innovation and category support activities to the categories. Although the retailers make their contributions in this area, it has been noted that they may choose not to promote their private label brands in certain categories (e.g. the milk category). Fifthly, the private label quality spectrum is largely similar across the categories, as a two-tier private label architecture is commonly used. The retailer has a big say in these policy decision areas that affect the coexistence of manufacturer brands and private label in FMCG/supermarket product categories. Therefore, with respect to research issue 6 (What is the nature of the coexistence relationship between private label and manufacturer in FMCG product categories and how is it driven?), the following research proposition arises:

Research Proposition 6: The mode of coexistence between private label and manufacturer brands expresses itself in the form of category-specific strategic management regimes driven by the retailer.

Role of Power in the Coexistence of Private Label and Manufacturer Brands

Analysis of Research Issue 7

Research Issue 7: What role is played by power in the mode of coexistence between private label and manufacturer brands in FMCG product categories?

The mode of coexistence between manufacturer brands and private label has been discussed in the preceding section, by category and by specific area of strategic policy. The analysis will now focus on the role played by power in the coexistence of the two types of brands. The research findings in each area of strategic policy are therefore restated in summary form and then further discussed using the bases of power (French and Raven 1959; Hunt 2015). While it is reasonable to say that there are power relationships between the two types of brands in the categories, the nature of the power relationships can only be accurately established through systematic analysis.

One might assume that in the New Zealand FMCG/supermarket environment characterised by high retail consolidation and concentration, manifested in the form of a duopoly, coercive power would be the dominant source of power employed by retailers in dealing with manufacturer brands, in relation to their own private label brands. Is this the situation on the ground? The analysis therefore additionally focuses on finding an answer to this question. It should also be noted that in this section, some of the research evidence (in the form of quotations) used already is reused in order to develop the points further by facilitating analysis from a different perspective: the perspective of power.

Power as it Relates to Private Label Growth and Equilibrium

Interpreting this from a power perspective, retailers own the retail shelves, which gives them and their private label legitimate power to make decisions regarding the growth of private label in relation to the manufacturer brand in the categories. The generally higher profit margins (for the retailers) on the part of private label brands compared to manufacturer brands make the option of higher retailer brand penetration attractive. The threat to continued category innovation and support comes about, though, because the manufacturer brands would then lack the resources and/or incentives to continue to innovate and support the categories. Retailers would therefore not want to bring about a situation where manufacturer brands would "withdraw from marketing, innovation" as that would be tantamount to "kill[ing] the category" if the "point where suppliers stop innovating" is reached. Facilitating manufacturer brand withdrawal from innovation and category support in this way can be

Table 8.10 Analysis of private label growth/share and equilibrium in the context of sources of power

Aspect of strategic policy	Research finding	Sample text evidence representing the views of most of the research interview participants
Private label growth and equilibrium	The research found that there is an equilibrium/end-point in private label growth/share, which represents an optimum balance between manufacturer brands and private label in the categories. The retailer would not wish to go beyond this point as this would be detrimental to the long-term strategic health of the respective categories from an innovation, category support and consumer choice point of view.	"If private label becomes too big, then the proprietary brand will withdraw from marketing, innovation [...]; and that's where it all comes from because private labels aren't innovators" (Interview QR1). With respect to retailer brand share growth "there is a point [...] where we can kill the category" (Interview R5a). Talking about the point beyond which retail chain ST would not want to go in terms of retailer brand growth/share in the five categories of milk, flour, cheese, breakfast cereals and tomato sauce: "The point where suppliers stop innovating" (Interview ST1).

Source: Prepared for this book from research interview data

interpreted as facilitating the withdrawal of direct and indirect rewards that the manufacturer brands bring to the categories for as long as they continue to get a fair share of the category. Reward power on the part of manufacturer brands therefore has a role to play in the manufacturer brand/retailer brand category relationship (Table 8.10).

In addition, the facts that private label brands "aren't innovators" as such, and that where "marketing, innovation [...] all comes from" is the manufacturer brand, put the manufacturer brand in the position of being the expert in innovation and category support in the category in relation to private label. Thus, the expert power wielded by manufacturer brands has relevance to the relationship between manufacturer brands and private label in the categories. In this way therefore, the dominant sources of power as far as this aspect of strategic policy is concerned are expert and reward power, and not coercive power on the part of retailers. Even if one wanted to interpret the reward power mentioned as being coercive power on the part of manufacturer brands, in the sense that retailers are threatened by withdrawal from innovation and category support, this would not be a coercive power base held by the retailer, who largely holds the balance of power, but rather held by the manufacturer in the form of countervailing power.

Power and Shelf/Merchandising Decisions

As far as retail shelves are concerned, the grocery retail chains are better off with the better-performing brands—in other words, for the sake of generating sales, profit and category growth. In the relationship between manufacturers and retailers, grocery retailers hold the balance of power due to factors such as high retail consolidation and concentration (Table 8.11).

They have the power to determine what goes on the shelves and what does not: "It's not the manufacturers that have the power. It's us that have the power. It's us that will determine how much of their product they are going to sell" (Interview T3). By inference, coercive power is an option open to the retailers to use if they wish. If the retailers did not want any of the manufacturers' brands/products to be on their shelves, the retailers would not stock them or would delist them. That would be fatal to any such brand/product in the New Zealand duopoly environment, at least with regard to the domestic market.

Research evidence has shown that the category management process allocates shelf space "on share" and other performance variables; for instance, the bigger the brand share, the bigger the space; the bigger the margin, the bigger the space. Prominent shelf positions such as eye-level space are also allocated according to the performance and contributions of the brand "to support" of the category. In addition, deletions usually target poorly performing brands/products, among other factors such as duplication. Even the non-performing retailer brand products "get deleted as everything else". In this regard, powerful manufacturer brands have a chance in the categories as such "strong brands [...], the power of the brand" give manufacturers countervailing power. Such powerful

Table 8.11 Analysis of shelf/merchandising decisions in the context of sources of power

Aspect of strategic policy	Research finding	Sample text evidence representing the views of most of the research interview participants
Shelf/merchandising decisions	The research found that category management decisions related to shelf space, shelf position and stocking/deletion/rationalisation of manufacturer brands and private label brands largely take into account the overall performance of the respective brands. There was little evidence of a clear-cut overexposure and/or overpush of private label brands, though there were isolated comments in connection with this. The isolated comments, however, were not corroborated by shelf data.	On shelf space and shelf position: Shelf space is allocated "On share basically; product stock turn [...]. If a product is high in margin but slower in sales, they might still get good shelf space" (Interview W2). "Each supermarket has a planogram person that dictates what stock is going to be sold. In terms of stock layout, it tends to be [that] obviously those top-selling brands tend to get the better shelf positions. So, for example, on shelf position, you probably get the top-selling brand in the middle, at eye level, and then you probably get your private label brand there as well because obviously they are trying to protect their own brand [...]. And it comes down to [the] support you are offering the retailers as well" (Interview W1). "Private label will always sit alongside its equivalent market leader in the fixture" (Interview QR1). On rationalisation and product deletion: "If we make a decision on a deletion for example, it's made because it's good for our consumers and it's good for our categories" (Interview QR1). "They do that with the house brand as well [...]; there [are] house brand products that aren't successful and they get deleted as everything else does" (Interview R1a). On stocking: "Ultimately, we have a lot of power in terms of [not stocking] products if we choose not to. But, if they have strong brands, they have some power [...], the power of the brand; so, we have to have them. So, the balance of power changes a little bit, depending on the strength of the brands basically" (Interview Q5a).

Source: Prepared for this book from research interview data

brands are usually the leading manufacturer brands or niche brands. In the rationalisation exercise, such brands survive, as indicated by the following comment: "A lot of retailers are now moving to either a two- or three-brand strategy across some of the categories anyway, and letting those bigger brands to do all the innovation" (Interview W1). The powerful brands are built and supported by resources and marketing expertise, thus expert power is seen to play a bigger role in the decisions related to the category management exercise, and not coercive power which would be possible due to the power imbalance between retailers and manufacturers. But of course, there were isolated incidents of research participants on the manufacturer side perceiving and mentioning some retailer actions as acts of coercion.

Additionally, the strategic policy of having the private label "sit alongside its equivalent market leader" on the display shelf is evidence of the private label brand drawing on the power of the manufacturer brand. This is a very common and significant aspect of the coexistence of the two types of brands in all the categories studied, and in both retail chains (and the supermarket groups) studied. Seeking to be displayed next to the strong manufacturer brands therefore can be interpreted as being driven by the referent power of the manufacturer brand. Therefore, referent power is important in the coexistence of the two types of brands. The private label brand simply wants to be associated with the powerful manufacturer brand(s) of each respective category as a way of more effectively physically positioning themselves for the customer/consumer.

With regard to the strategic policy area of shelf/merchandising decisions relating to manufacturer brands and retailer brands, expert and referent sources of power are seen to play a more active role compared to coercive power, which one would have expected to play a bigger role under the given circumstances.

Power and Category Management Arrangements

As the owners of the grocery retail shelves, the retailers have the prerogative of appointing to CC the manufacturer/supplier "who [they] want to do it". The power to appoint who they want is more a reflection of their legitimate ownership of the retail outlets than anything else. The retailer's basis of power here can be interpreted as legitimate power. Although it has been established that the appointment of CCs can be based on other factors, inclusive of political factors, the research found that the most common pattern was that of appointing "the market leader". Reasons for appointing the market leader include: they have a large share of the category and they are resourceful enough to spend on the category; and they are in a leadership position because "they are good" and "they execute well", including execution in the areas of market research, planning, branding and advertising, promotions and so on. The retailer therefore would usually appoint the expert in the category. A reasonable interpretation of this state of affairs would be that the appointment of the CC is largely driven by the expert power the manufacturer/supplier wields, as well as the referent power of the leader's brands. In addition, the CC's ploughing back of resources to manage the category could be interpreted as an act of reward to the category and the retailer, for being allowed to take up such a role. So, there is also an element of reward power (Table 8.12).

Power and Driving Category Growth through Innovation and Category Support

Manufacturer brands "have the technology and knowledge" to "drive innovation" in the categories, and private labels are largely dependent on this knowledge and technology, which can itself be described as an area of expertise. Thus, expert power on the part of manufacturer brands plays an important role in the coexistence of the two types of brands in the categories. With the manufacturer brands as the category leaders on innovation and category support, and the retailer brand as "a follower", the expert power of the manufacturer brand gets further defined in the relationship. Manufacturer brands also spend a lot of resources such as "advertising revenue and cooperative dollars" to "drive promotional programmes", thereby consolidating this leadership role. Strategic policies governing the coexistence of the two types of brands

Table 8.12 Analysis of category management arrangements in the context of sources of power

Aspect of strategic policy	Research finding	Sample text evidence representing the views of most of the research interview participants
Category management arrangements	The research found that the retailer has full responsibility and power to appoint the CC(s), and the retailer has the final say on category matters; thus, the retailer maintains strategic control. In most cases, however, it is the leading manufacturer brand in the category whose company gets to be appointed CC.	"The market leader in that category would do the relay, but [...] we have the final say on what happens [...]. We look at the category and say for example if it's cheese; [name of manufacturer] make all the cheese [...] they have got 75% of the cheese, so they [...] do the relay for us [...]. [And, with regard to the appointment of category champions] we appoint who we want to do it [...]. We generally look for a couple of things; how they support our store as a company, how big they are in the category, and [...] their rep, the money they spend with us and the promotions they give us, TV spend [...]. We go after market leaders, and generally if they are the market leader, there is a reason why they are; because they are good at store level; they execute well" (Interview R4).

Source: Prepared for this study from research interview data

in the categories, which are largely driven by the legitimate owner of the retail shelves (the retailer), generally seek to accommodate manufacturers who are experts in this area and who have the necessary resources as well (Table 8.13).

There is no direct instruction from the retailer that the manufacturer should drive innovation and category support, but it appears that there is tacit understanding in the industry. Furthermore, naturally, manufacturers seek to outcompete other manufacturers in the categories through innovation, category support and satisfying the consumer. Despite the retailers having the balance of power and legitimate ownership of the point of sale, it is rather the expert power of manufacturer brands and

Table 8.13 Analysis of driving category growth through innovation and category support in the context of sources of power

Aspect of strategic policy	Research finding	Sample text evidence representing the views of most of the research interview participants
Driving category growth through innovation and category support	The research found that manufacturer brands have superior collective capacity for innovation and category support compared to retailer brands, and that it is the manufacturer brands that do the bulk of driving the categories through innovation and marketing activities. It is also generally regarded as the manufacturer's area of responsibility to bring about innovation and category support. Manufacturer contributions towards innovation and category support, alongside consumer choice, have relevance for strategic policies that govern the balance and coexistence between the two types of brands in the categories.	"You need that branded product [manufacturer brand] to drive innovation. You need that branded product to drive promotional programmes. You need the branded product to generate advertising revenue and cooperative dollars [...], all those sorts of things" (Interview QR3). "Manufacturer brands drive the category, bring innovation and generally drive promotion. So, it's not the role of store brands. Store brands are a follower" (Interview W2). In the area of product innovation, retailers are largely dependent on suppliers "because they [suppliers] have the technology and the knowledge" (Interview Y1).

Source: Prepared for this book from research interview data

the recognition of this by the retailers that largely governs the coexistence of the two types of brands in the categories on this aspect of strategic policy. Even consumer loyalty to certain brands can be interpreted as a result of the resources and expertise that have gone into developing those brands. The dominant source of power, therefore, regarding the coexistence of manufacturer and retailer brands as far as this area of strategic policy is concerned, is expert power. In addition, French and Raven's (1959) typology does not adequately cater for the total picture. Manufacturer brands' capacity to deliver product innovation and category support is mainly due to their resources, therefore resources as a form of power Mintzberg (1983) additionally explain the power relationship with respect to innovation and category support.

Power and Private Label Quality Spectrum

The full quality spectrum in any category would normally consist of three price–quality tiers; economy brands, mid-segment (standard) brands and premium brands. The top tier, which is the premium segment, is not fully represented in the private label portfolio of New Zealand grocery retail chains. The mainstream private label brands that straddle the mid-segment and premium (i.e. mid-to-premium) markets cannot really be regarded as top-tier brands. In this regard, despite statements such as "we don't think the market is big enough for three [tiers]", it is interesting to note that the third tier (top tier) is catered for by manufacturer brands: "you have got up here [upper segment/top tier] your market leaders [manufacturer brands]". Some of the key requirements for success of these top tier brands include brand development, brand management, marketing support (largely advertising) and continuous product and brand innovation (Table 8.14).

Private label brands would hardly want to commit resources to such an intensive marketing programme to build and support top-tier brands. In addition, it is not just a matter of having a good-quality product, as an external company can easily produce this for the retailer brand, but rather a matter of having the expertise and resources to run an intensive marketing programme consistent with the level of branding required for such top-tier products. Therefore, to do that for little return, given the small size of the New Zealand market, would not make economic sense. The retailers rely on the manufacturer brands' expertise and resources to fill the gap that is missing in their retailer brand portfolio. Currently, therefore, private label strategy takes into account the expert power of leading manufacturer brands to cater for the gaps in the private label architecture. Additionally, as has already been noted, the mainstream private label brands get displayed right next to leading manufacturer brands, a practice that was interpreted in this book as being driven by manufacturer brand referent power.

Table 8.14 Analysis of private label quality spectrum in the context of sources of power

Aspect of strategic policy	Research finding	Sample text evidence representing the views of most of the research interview participants
Private label quality spectrum	The research found that the retailers employ a two-tier private label portfolio consisting of an economy and a middle/standard-to-premium brand. It is the top-tier, premium brand that is absent from their private label portfolio.	"[Name of private label] is your economy [brand]. [Name of private label]; that will be perceived as your middle row, and then you have got up here [upper segment/top tier] your market leaders [manufacturer brands]" (Interview Q4). "We will remain on two [tiers]. We don't think the market is big enough for three" (Interview QR3). On the possibility of a three-tier retailer brand system: "I wouldn't be surprised, but I think we have the disadvantage of being such a small market [New Zealand]. It's going to be more difficult to get the separation between them" (Interview W10).

Source: Prepared for this study from research interview data

Wrap-up of Discussion on Research Issue 7

Strategic policy aspects of the coexistence of manufacturer brands and private label in FMCG/supermarket product categories were explored in this section, covering the areas of private label growth and equilibrium, shelf/merchandising decisions, category management arrangements, driving category growth through innovation and category support and private label quality spectrum, and using the bases of power as an interpretive framework. Despite the high retail consolidation and concentration in the New Zealand grocery retail industry and the power imbalance in favour of the grocery retail chains in relation to manufacturers/suppliers, and contrary to expectations, coercive power was not found to be the dominant source of power governing the coexistence of the two types of brands. Indeed, the coexistence and related equilibrium are largely driven by non-coercive sources of power, namely expert and referent power (Chimhundu 2016). With respect to research issue 7, therefore (i.e. What role is played by power in the mode of coexistence between private label and manufacturer brands in FMCG product categories?), the following research propositions arise:

Research Proposition 7a: The mode of coexistence between private label and manufacturer in FMCG product categories is rooted in the theory of power.

Research Proposition 7b: The mode of coexistence between private label and manufacturer in FMCG product categories is largely driven by expert and referent bases of power rather than coercive power.

Chapter Recap

This chapter has restated the research issues, provided case summaries of FMCG categories and grocery retailers, and analysed the research issues. The balance between private label and manufacturer brands, comparative capacity for product innovation and category support, consumer choice and product category strategic policies were examined. Some of the interview quotations integrated into the analysis of research issues in this chapter were previously published by this author. The final chapter of this book will address the conclusions and implications of this study.

References

ACNielsen. (2005). *The power of private label: A review of growth trends around the world.* New York, NY: ACNielsen.

Anonymous. (2005). House brand strategy doesn't quite check out. *The Age.* Retrieved December 3, 2017, from http://www.theage.com.au/news/Business/House-brand-strategy-doesnt-quite-check-out/2005/04/01/1112302232004.html

Chimhundu, R. (2012). Marketing manufacturer and retailer house brands: A study of shelf space in two product categories. *International Journal of Marketing Studies, 4*(6), 35–43.

Chimhundu, R. (2016). Marketing store brands and manufacturer brands: Role of referent and expert power in merchandising decisions. *Journal of Brand Management, 23*(5), 24–40.

Chimhundu, R., & Chadee, D. (2013). Private label brand trends in the grocery retail industry. *Journal of Euromarketing, 22*(1), 36–47.

Chimhundu, R., Hamlin, R. P., & McNeill, L. (2010). Impact of manufacturer brand innovation on retailer brands. *International Journal of Business and Management, 5*(9), 10–18.

Chimhundu, R., Hamlin, R. P., & McNeill, L. (2011). Retailer brand share statistics in four developed economies from 1992 to 2005: Some observations and implications. *British Food Journal, 113*(3), 391–403.

Chimhundu, R., Kong, E., & Gururajan, R. (2015). Category captain arrangements in grocery retail marketing. *Asia Pacific Journal of Marketing and Logistics, 27*(3), 368–384.

French, J. R. P., & Raven, B. (1959). The bases of social power. In D. Cartwright (Ed.), *Studies in social power*. Institute of Social Research (pp. 150–167), Ann Arbor, MI:The University of Michigan.

Hunt, S. D. (2015). The bases of power approach to channel relationships: Has marketing's scholarship been misguided? *Journal of Marketing Management, 31*(7–8), 747–764.

Mintzberg, H. (1983). *Power in and around organisations*. Englewood Cliffs, NJ: Prentice Hall.

9

Conclusions and Implications of this Book

Overview

This chapter discusses the conclusions and implications of this study. It outlines a summary of outcomes of the research issues and goes on to address specific interpretations. Answers are provided to the research sub-questions of this book. Specifically, a summary is given of the outcomes of the research issues, addressing the findings of the study in the context of the literature, the modified conceptual framework, the theoretical implications, the implications for marketing practice, conclusions that may be drawn from the book and directions for further research.

Summary of Outcomes on the Research Issues

Based on the analysis of the research issues as extensively discussed in the three previous chapters, Table 9.1 gives a summary and concluding snapshot of the research outcomes for each of the research issues.

A discussion of the research findings against key aspects of the literature will now follow.

R. Chimhundu, *Marketing Food Brands*,
https://doi.org/10.1007/978-3-319-75832-9_9

Table 9.1 Outcomes on the research issues analysed

Specific research issues (i.e. sub-questions)		Arising research propositions
Research Issue 1	What is the general nature of the state of balance between private label and manufacturer brands in FMCG product categories?	*Research Proposition 1a:* Equilibrium points exist in the aggregate long-term share trends of private label and manufacturer brands in FMCG product categories. *Research Proposition 1b:* There are two possible states of equilibrium, premature and mature equilibrium, in the long-term share trends of private label and manufacturer brands in FMCG product categories.
Research Issue 2a	In a grocery retail landscape characterised by high retail concentration, what is the nature of the state of balance between private label and manufacturer brands in FMCG product categories?	*Research Proposition 2a:* In a grocery retail landscape characterised by high retail concentration, there is an existence of equilibrium points between private label and manufacturer brands in FMCG product categories.
Research Issue 2b	How does the balance between private label and manufacturer brands in FMCG product categories compare by category?	*Research Proposition 2b:* The balance between private label and manufacturer brands in FMCG product categories differs from category to category.
Research Issue 2c	How does the balance between private label and manufacturer brands in FMCG product categories compare between grocery retailers?	*Research Proposition 2c:* The balance between private label and manufacturer brands in FMCG product categories differs between grocery retailers but follows a similar pattern across similar product categories of the different retailers.
Research Issue 3a	How do private label and manufacturer brands compare on capacity to innovate in the FMCG product categories?	*Research Proposition 3a:* Due to their greater resources, manufacturer brands have superior collective capacity to innovate at a higher rate than private label.
Research Issue 3b	How do private label and manufacturer brands compare on capacity to contribute to product category marketing support and development?	*Research Proposition 3b:* Due to their resources, manufacturer brands have superior collective capacity to make a greater contribution to category marketing support and development than private label.

(*continued*)

Table 9.1 (continued)

Specific research issues (i.e. sub-questions)		Arising research propositions
Research Issue 4a	What is the state of awareness of FMCG retailers on the comparative capacity of private label and manufacturer brands to innovate and give marketing support to the product categories?	*Research Proposition 4a:* Retailers are aware of manufacturer brands' superior capacity for innovation and category development.
Research Issue 4b	What is the strategic stance of the FMCG retailers in relation to this comparative capacity to innovate and give marketing support to the product categories?	*Research Proposition 4b:* Retailers seek to draw upon this capacity as part of a coherent, pooled interdependence strategy.
Research Issue 4c	How is this strategic stance related to policies on the coexistence of private label and manufacturer brands in FMCG product categories?	*Research Proposition 4c:* This aspect of strategic dependency has relevance for the determination of policies that govern the coexistence of private label and manufacturer brands in FMCG product categories.
Research Issue 5	What role is played by consumer choice in shaping the coexistence of private label and manufacturer brands in FMCG product categories?	*Research Proposition* 5: Consumer choice considerations have relevance for the determination of policies that govern the coexistence of private label and manufacturer brands in FMCG product categories.
Research Issue 6	What is the nature of the coexistence relationship between private label and manufacturer brands in FMCG product categories and how is it driven?	*Research Proposition* 6: The mode of coexistence between private label and manufacturer brands expresses itself in the form of category-specific strategic management regimes driven by the retailer.
Research Issue 7	What role is played by power in the mode of coexistence between private label and manufacturer brands in FMCG product categories?	*Research Proposition 7a:* The mode of coexistence between private label manufacturer brands in FMCG product categories is rooted in the theory of power. *Research Proposition 7b:* The mode of coexistence between private label and manufacturer brands in FMCG product categories is largely driven by expert and referent bases of power rather than coercive power.

Source: Prepared for this book based on research evidence

Addressing the Findings of the Book in the Context of the Literature

This section is divided into three themes: the balance between private label and manufacturer brands in FMCG/supermarket product categories; product innovation, category development/category support and consumer choice; and strategic policies on the coexistence of private label and manufacturer brands in the categories. The approach to this section is that the research propositions outlined in the "Summary of Outcomes for Research Issues" are discussed in the context of the literature. Some of the propositions are handled jointly where there is overlap. Additionally, the approach is that the research propositions are stated first, acting as sub-headings for the discussion that follows them.

Balancing Private Label and Manufacturer Brands in FMCG/Supermarket Product Categories

Concluding comments on research propositions 1a, 1b and 2a:

Research Proposition 1a: Equilibrium points exist in the aggregate long-term share trends of private label and manufacturer brands in FMCG product categories.

Research Proposition 1b: There are two possible states of equilibrium, premature and mature equilibrium, in the long-term share trends of private label and manufacturer brands in FMCG product categories.

Research Proposition 2a: In a grocery retail landscape characterised by high retail concentration, there is an existence of equilibrium points between private label and manufacturer brands in FMCG product categories.

The literature review has established that the FMCG/supermarket environment has generally become more consolidated and concentrated, especially in some economies; and that grocery retail consolidation/concentration has been linked with pushing private label share to high levels

(Burt 2000; Coriolis Research 2002; Cotterill 1997; Defra 2006; Hollingsworth 2004; Nielsen 2014; Rizkallah and Miller 2015). In the manufacturer–retailer relationship, FMCG retailers largely hold the balance of power (Berthon et al. 1997; Hogarth-Scott 1999; Hollingsworth 2004; Hovhannisyan and Bozic 2013, 2016; Stanković and Končar 2014; Sutton-Brady et al. 2015; Panigyrakis and Veloutsou 2000; Weitz and Wang 2004) and the powerful retailers have become gatekeepers of the grocery retail shelves (ACNielsen et al. 2006). It is reasoned that this situation would predispose the retailers to want to capitalise on their power to push their private label share to high levels, as a way of exploiting private label's higher capacity to generate profit (ACNielsen 2005; Burt 2000; Burt and Sparks 2003; Coriolis Research 2002; Cotterill 1997; Defra 2006; Galbraith 1952; Porter 1976; Rizkallah and Miller 2015), and the systematic push was reasonably perceived as having no end in sight, leading to an overdominance of private label in the categories. Two interpretations have come out of this issue, representing a contradiction in the literature (Chimhundu et al. 2011). One line of reasoning is that in highly concentrated grocery retail environments there is bound to be an overdominance of private label. This line of reasoning is consistent with an adversarial and aggressive approach to private label growth on the part of retail chains. The other line of reasoning is that there is a high level of strategic dependency between private label and manufacturer brands, and therefore private label share and related merchandising measures will not go beyond a level that would jeopardise manufacturer brand contributions to the categories.

The results of the preliminary study on private label and manufacturer brand share trends did not support a relentless advance in private label share growth, and the results are not consistent with the adversarial and aggressive approach of pushing private label to overdominance, despite the environment of high retail consolidation/concentration and retailer power. The results indicate that there are equilibrium points in grocery product categories between private label and manufacturer brands that exist independently of the level of power enjoyed by the retailer, even in situations where the power of the retailer would theoretically allow them to advance this penetration to very much higher levels.

Additionally, the intensive primary study carried out in the New Zealand FMCG/supermarket industry established that despite the private label ambitions of the two major retail chains (in the New Zealand duopoly), which are both geared towards increasing private label penetration from current levels, there is still an end in sight on the part of both retail chains: a point beyond which the private label is not desired to go, which is the point of ultimate equilibrium. This gives support to the argument that a 'strategy' of some sort does exist, which is generally understood, especially in the hierarchy of the retail chains, and is a systematically applied, high level (but unreported) retailer strategy. Additionally, it was found that this is not necessarily explicitly communicated with the participating manufacturers. One might therefore argue that there is an element of tacit understanding in the industry on the matter. Despite the shift in the balance of power to retailers, which may mean that the retailers largely have the final say, it appears that the actual final say is determined through rational business judgement, as the retailers would not be willing to pursue actions that would be detrimental to the categories. This reasoning is consistent with aspects of the literature that have suggested that "the fact that private labels have low share in a category does not imply that a particular retailer cannot create a successful program in that category" (Hoch and Banerji 1993: 66). This particular point can even be extended to argue that theoretically, the retailers have the capability to be overdominant in some categories if they wanted it that way. Therefore, deciding whether to create the programme, and with what objectives in terms of the size of the private label in the categories in relation to manufacturer brands, is a matter of rational business judgement.

Propositions 1 and 2a are therefore given support by the results of this study and the subsequent discussion, and one of the explanations on the existence of equilibrium points is that retailer power is largely used in the context of what is good for the categories long term, and not necessarily in the context of the maximisation of retailer benefits at the expense of the participating manufacturers, as this would not be sustainable. This reasoning is further extended to give rise to subsequent propositions relating to the contributions of private label and manufacturer brands towards innovation and the development of the categories, as well as to the bases of power that are dominant in the coexistence of the two types of brands in FMCG/supermarket product categories.

Concluding Comments on Research Propositions 2b and 2c

Research Proposition 2b: The balance between private label and manufacturer brands in FMCG product categories differs from category to category.

Research Proposition 2c: The balance between private label and manufacturer brands in FMCG product categories differs between grocery retailers but follows a similar pattern across similar product categories of the different retailers.

Private label share is not the same across categories (ACNielsen 2005; Hoch and Banerji 1993; Nielsen 2014; Sirimanne 2016; Steenkamp and Dekimpe 1997), and FMCG/supermarket product categories also differ on a number of characteristics that include size, range of competing brands and products (ACNielsen 2005), rate of category innovation (Coriolis Research 2002), level of technology (Lehmann and Winer 2002), level of category commoditisation and related variables. It was therefore reasoned that the product categories offer different opportunities and challenges to the retailers. Again, since certain categories such as those that are commoditised are prone to high private label penetration in relation to others, retailer strategic objectives and policies are expected to differ across the categories.

A comparison of the food categories studied (milk, flour, cheese, breakfast cereals and tomato sauce) has shown that there are differences in private label penetration as expected, and retailer strategic objectives and policies differ as well across the categories. Categories such as milk and flour have high private label penetration, while categories such as breakfast cereals and tomato sauce have relatively low private label penetration. The aggregate private label shares discussed in the preliminary study are therefore made up of different categories that have different degrees of balance between private label and manufacturer brands. Differences have also been noted in a number of other aspects studied, such as shelf space and facings, shelf position, number of brands and number of products in the categories. This state of affairs therefore gives support to research proposition 2b and makes the research proposition relevant. In addition, although shelf and related measures were found to be different in the two major retail chains and their respective supermarkets, there is a similar pattern between the retail chains and among the supermarket groups, where similar categories have high levels of private label penetration, e.g. milk and flour in both retail chains QR and ST. Similar categories have low private label penetration (e.g. breakfast cereals and tomato sauce), or may have no private label (e.g. deodorants and moisturisers). This is a response to the opportunities available in each category relating to private label, and these opportunities are discerned by both retail chains. In this respect, support is given to research proposition 2c as a relevant proposition.

On the shelves, there was a mix of results on shelf measures for the two types of brands. There were instances where either the private label or the manufacturer brand were judged to be occupying either more than, or less than, or the appropriate amount of space and prime shelf position. The mix of these occurrences was such that the results of the study are not consistent with a private label phenomenon that has become overdominant, not only from the viewpoint of share/penetration, but also from the perspective of merchandising measures. In most categories, manufacturer brands still dominate. Again, while the nature of the category may partly determine the nature of coexistence, one may still reason that the differences between what the categories have to offer to the retailers also mean that retailer category strategic management regimes are likely to differ by category; an issue that is examined in a later proposition. By implication, this book further argues that it is not just market forces that determine the balance in the coexistence of private label and manufacturer brands in FMCG/supermarket product categories, but that retailer strategic objectives regarding private label have a role in it as well.

Product Innovation, Category Marketing Support and Consumer Choice

Concluding Comments on Research Propositions 3a and 3b

Research Proposition 3a: Due to their resources, manufacturer brands have superior collective capacity to innovate at a higher rate than private label.

Research Proposition 3b: Due to their resources, manufacturer brands have superior collective capacity to make a greater contribution to category marketing support and development than private label.

It was established in the literature review that both private label and manufacturer brands engage in product innovation and category support activities. Researchers, however, hold different views on the state of innovation and the state of contribution to the development of categories by the two types of brands. According to Conn (2005), one perspective is that private label brands are leading the way and setting the standards on innovation and retailers have been seen to be boosting their capacity for innovation (Lindsay 2004), in addition to private label in certain countries having moved away from copying competition to setting its own trends (Silverman 2004). Thus, private label brands have become masters of their own destinies on innovation. The contrasting view notes that historically, retailers have largely been followers of manufacturer brands on innovation (Aribarg et al. 2014; Coelho do Vale and Verga-Matos 2015; Hoch and Banerji 1993; Olbrich et al. 2016). Generally, private label development is not backed by enough research and development money and cannot afford the necessary resources (Conn 2005; Steiner 2004). In addition, supermarket categories would naturally consist of many specialised manufacturers whose manufacturer brands, collectively, are expected to have more resources, skills and knowledge in the areas of product innovation and category support than private label brands. There are also two perspectives in the literature on the extent to which private label can actually develop the categories. One view, consistent with research carried out by Putsis and Dhar (1996) is that the private label is capable of expanding category expenditure and developing the market, and not just stealing share from manufacturer brands. A contrary view, however (Anonymous 2005), is that it is the manufacturer brand that actually expands the categories.

The results of the study, in relation to capacity for innovation and rate of innovation, have shown that the resources (and expertise) of manufacturers enable them to collectively contribute more towards product innovation in the categories than private label. Research proposition 3a therefore reflects the research evidence. In this proposition, the term "resources" incorporates not only facilities and finance, but also expertise from the viewpoint of expert resources. The results of the study as far as category support is concerned have shown that manufacturer brands do the bulk of the category development activities. The collective resources of the different manufacturers/suppliers that are ploughed back into the categories through category marketing support are greater in magnitude than those of private label. Therefore, in the same way as research proposition 3a, research proposition 3b is judged to be relevant and is given support.

While the results of the study are very much in line with Hoch and Banerji (1993) and Steiner (2004) concerning capacity for innovation, and with Anonymous (2005) on the issue of capacity for category support, it should be noted that the results are specific to the New Zealand FMCG/supermarket industry, and on the global level, to FMCG retail environments that are characterised by high retail concentration and that are still at the stage of developing private label to higher levels. From a private label perspective, the New Zealand private label industry is still developing, unlike some of the more mature private label industries such as the UK and Switzerland. In this regard, it can be suggested that, on the matter of private label and manufacturer brand contributions to innovation and category support, there is a bigger gap between the contributions of the two types of brands to FMCG/supermarket categories in New Zealand than, say, in the UK. It can equally be reasoned that since private label development is at different stages in different parts of the world, the situation in terms of private label and manufacturer brand contributions to product innovation and category support is economy- and industry-specific. In industries that have yet to develop their private label portfolios to the fullest, manufacturer brands are expected to do much more in the areas of innovation and category support than private label. The study has shown that the New Zealand FMCG/supermarket environment is one such industry, and on a global scale, industries with similar conditions are expected to have similar characteristics.

Concluding Comments on Research Propositions 4a, 4b and 4c

Research Proposition 4a: Retailers are aware of manufacturer brands' superior capacity for innovation and category development.

Research Proposition 4b: Retailers seek to draw upon this capacity as part of a coherent, pooled interdependence strategy.

Research Proposition 4c: This aspect of strategic dependency has relevance for the determination of policies that govern the coexistence of private label and manufacturer brands in FMCG product categories.

Since retailers are actively involved in the management of FMCG/supermarket product categories, it is to be expected that they would be very much aware of the state of affairs concerning private label and manufacturer brand capacity for innovation and category support in the product categories, in the same way as they would be aware of product and category trends relating to sales and profit performance.

Furthermore, the literature review has established that researchers are generally in agreement on the importance of innovation as a driving force in the growth of companies and categories (e.g. Anonymous 2004; Brenner 1994; BCG 2005; Doyle and Bridgewater 1998; Guinet and Pilat 1999; Hardaker 1998; Kung and Schmid 2015; Robert 1995). Research has also demonstrated that more innovative categories tend to be more successful than less innovative ones (Booz et al. 1982). Brand building and differentiation through a variety of marketing activities, including advertising, are seen as essential ingredients of the continued development of product categories.

The consumer packaged goods literature has, however, largely portrayed manufacturer brand innovation in relation to private label as a competitive tool that is employed against private label, in addition to competing with other manufacturer brands (e.g. Coriolis Research 2002; Information Resources Inc. 2005; Kumar and Steenkamp 2007b; Verhoef et al. 2002). It largely suggests that manufacturer brand innovation is inhibiting private label in FMCG categories. It was noted that the alternative view of manufacturer brand innovation as an enhancer of private label has not been investigated in depth and has not been given prominence in the mainstream academic literature (Chimhundu et al. 2010). For instance, the imitative and parasitic behaviour of private label (e.g. Coelho do Vale and Verga-Matos 2015; Collins-Dodd and Zaichkowsky 1999; Harvey 2000; Hoch and Banerji 1993; Ogbonna and Wilkinson 1998; Olbrich et al. 2016) has been reported in the literature, and one would imagine that retailers are positively benefiting from manufacturer brand innovation in that way and in related ways. It can therefore be reasoned that if there are aspects of manufacturer brand innovation that enhance private label, those aspects would most likely be taken into account by the powerful retail chains in the determination of the policies and strategies that govern the coexistence of their private labels with manufacturer brands in the product categories. One might further reason that manufacturer brands that are more innovative and supportive to the categories are viewed more positively by grocery retailers than those that are not because the grocery retailers benefit more from them.

The results of this research are therefore in line with research propositions 4a and 4b respectively. In the process, it is shown that there is an element of contributing to a pool from which the other party draws. In addition, four themes were drawn from the discussions on the topic of how good or bad manufacturer brand innovation activities are for private label. The study has shown the following: the competitiveness of innovative manufacturer brands inhibits private label (theme 1); private label is enhanced by manufacturer brand innovation driving the growth of products (theme 2); the adaptation of successful manufacturer brand innovations by private label enhances the private label (theme 3); and the customer pulling power of innovative manufacturer brands enhances private label (theme 4). These are areas of strategic dependency between manufacturer brands and private label. Although the competitiveness of innovative manufacturer brands inhibits private label, manufacturer brand innovation and category support actually have a huge positive impact on the welfare of private label in FMCG/supermarket categories on aspects related to driving category growth, private label adapting successful manufacturer brand innovations, and private label tapping into the customer pulling power of innovative manufacturer brands. These three aspects therefore make research proposition 4b relevant.

This book argues that strategic policies that govern the coexistence of private label and manufacturer brands in the categories are partly influenced by retailer perceptions of manufacturer brands' ability to enhance the private label (Chimhundu et al. 2010). Manufacturer brand innovation and category support largely have a positive impact on private label in grocery retail categories, and private label is largely strategically dependent on manufacturer brands in these areas. This aspect of strategic dependency therefore has relevance for the determination of policies that govern the coexistence of private label and manufacturer brands in FMCG/supermarket product categories, and research proposition 4c becomes relevant in this regard as it is given support by the results and by this discussion. Going beyond the superficial level of competition between the two types of brands and demonstrating the converse aspects of the systematic enhancement of private label is in itself a step forward in better understanding the coexistence of private label and manufacturer brands in the categories. Aspects of strategic dependency play a key role.

Concluding Comments on Research Proposition 5

Research Proposition 5: Consumer choice considerations have relevance for the determination of policies that govern the coexistence of private label and manufacturer brands in FMCG product categories.

From the literature, it was noted that the consumer is the ultimate customer for both private label and manufacturer brands. The consumer and related consumer choice considerations are relevant to the coexistence of private label and manufacturer brands in FMCG/supermarket product categories. This is in line with the reviewed definition of category management that includes consumer focus as a key element. Consumers drive what goes on in the categories (ACNielsen et al. 2006), alongside other factors. Consumers have the freedom to choose among competing offerings (Kaswengi and Diallo 2015; Nelson 2002; Olbrich et al. 2016) and they want brand/product selection as well. It was further established in the review of the literature that in the category management set-up, the private label is seen as being protected by the retailer (e.g. Major and McTaggart 2005) and has the privilege of being in control of its own marketing mix and that of competitor brands (Hoch et al. 2002a, b). In addition, retailers have their own strategic objectives for private label and can employ specific strategic management regimes at their discretion.

The results of the study are in line with research proposition 5 as a relevant proposition. Consumer choice considerations have a bearing on the determination of policies that govern the coexistence of private label and manufacturer brands in FMCG/supermarket product categories. The consumer has not been lost in the equation, despite the fact that the retailers may have to compromise consumer issues to advance certain strategic objectives in terms of private label. In the highly consolidated and concentrated grocery retail environment where retailers have become increasingly powerful, while the retailers have other strategic interests with respect to their private label (e.g. rationalising the categories and accommodating the private label) and such considerations may have compromised consumer choice issues to some degree, consumer choice is still a valid concern when determining the balance between private label and manufacturer brands in the categories. Private label overdominance would compromise consumer focus to the extent that the business viability of the categories would be threatened. Therefore, although the retailers, as owners of the private label, hold the balance of power in the relationship with manufacturers as owners of the manufacturer brands, consumers still have a measure of power, even in the highly concentrated grocery retail environment, as they continue to demand manufacturer brands. This tends to give the manufacturer brand a measure of power as well.

Category Strategic Policies on the Coexistence of Private Label and Manufacturer Brands in FMCG/ Supermarket Product Categories

Concluding Comments on Research Proposition 6

Research Proposition 6: The mode of coexistence between private label and manufacturer brands expresses itself in the form of category-specific strategic management regimes driven by the retailer.

As noted in the discussion of the preceding propositions, the mode of coexistence between private label and manufacturer brands takes into account areas of strategic dependency between the two types of brands, as well as consumer choice. Due to the fact that it is widely recognised in the literature that power in the FMCG sector has shifted from manufacturers to retailers, and that the balance of power is in the hands of the retailers (ACNielsen et al. 2006; Berthon et al. 1997; Hogarth-Scott 1999; Hollingsworth 2004; Hovhannisyan and Bozic 2013, 2016; Stanković and Končar 2014; Sutton-Brady et al. 2015; Kumar and Steenkamp 2007a, b; Panigyrakis and Veloutsou 2000; Weitz and Wang 2004), retailers are expected to have a bigger say than manufacturers on how private label and manufacturer brands should coexist in the grocery retail categories in the matters of strategic dependency, consumer choice, merchandising issues, private label growth and share and so on. The study has shown that despite the fact that the five categories, milk, flour, cheese, breakfast cereals and tomato sauce, are managed with the help of participating manufacturers, especially the leading manufacturers, and with varying degrees of participation depending on the retail chain and the category, the retailers are actually very much in charge of what goes on in all the categories studied and in both retail chains (and all four supermarket groups studied). In the context of the literature (e.g. Kurtulus and Toktay 2005) which has hinted that "retailers who become too dependent on their category captains risk a strategic loss of power" (p. 59), this book has shown that retailers have the final say on policies and strategies relating to the coexistence of private label and manufacturer brands in the categories. There is no evidence of any strategic loss of power on the part of the retailers in terms of category management arrangements. This is in line with research proposition 6 on retailers driving the strategic management regimes. It is therefore largely retailer strategic thinking that drives private label and manufacturer brand coexistence in the categories.

The literature has also established that there are inherent differences in the categories (ACNielsen 2005; Coriolis Research 2002; Hoch and Banerji 1993; Lehmann and Winer 2002) and that the categories offer different opportunities and challenges for the private label brands, therefore the retailers would most likely equally put in place policies that are category-specific. The study found that retailer expectations of the coexistence of the two types of brands in the categories are not the same for milk, flour, cheese, breakfast cereals and tomato sauce. Retailer strategic objectives regarding the level to which the private label should grow differ across the categories. The other part of proposition 6 that relates to category-specific strategic management regimes is also in line with the results of this study. It can be further reasoned that due to their control, the retailers can influence to some extent the aspects that determine the final composition of private label and manufacturer brands participating in the categories, although market forces are also at play. In addition, it was found that across the studied categories, the retailers expect that manufacturer brands do the bulk of driving the categories through innovation and category support. Although retailers make their contributions in this regard, it was noted that the retailers promote their private label in some categories but choose not to do so in others (e.g. the milk category), and these are category-specific decisions. As far as private label portfolio is concerned, it was found that the retailers have adopted a two-tier quality spectrum across the five categories and two retail chains. In the context of the four-generation private label typology (Laaksonen 1994; Laaksonen and Reynolds 1994), it is largely the fourth-generation private label that is missing from the private label portfolio, but the fourth generation does exist on the manufacturer brand side. A full private label quality spectrum is not being employed, and discussions held with the research participants confirmed that this is the model of coexistence currently in place.

It could be argued, on the other hand, that while retailers may not be able to conduct innovation and category support to the same level as manufacturers in all categories, they would definitely be able to do so in some (at least a few) categories if they wanted to. This reasoning is consistent with the argument put forward by Hoch and Banerji (1993), that "the fact that private labels have low share in a category does not imply that a particular retailer cannot create a successful program in that category" (p. 66). Innovation related to fourth-generation private label and category support linked to brand advertising would form a good part of that "successful programme". In this regard, retailer strategic management regimes in the different categories are tuned to the level of "success" that the retailers want to achieve in a category. Therefore, both strategic management regimes (designed to achieve certain outcomes with respect to the coexistence of private label and manufacturer brands) and consumer considerations have a part to play in the coexistence.

Concluding Comments on Research Propositions 7a and 7b

Research Proposition 7a: The mode of coexistence between private label and manufacturer brands in FMCG product categories is rooted in the theory of power.

Research Proposition 7b: The mode of coexistence between private label and manufacturer brands in FMCG product categories is largely driven by expert and referent bases of power rather than coercive power.

The radically altered FMCG landscape (Kumar and Steenkamp 2007a) is characterised by increased retailer power (Hogarth-Scott 1999; Hollingsworth 2004; Hovhannisyan and Bozic 2013, 2016; Nielsen 2014; Rizkallah and Miller 2015; Stanković and Končar 2014) and direct competition (Steenkamp et al. 2010) between brands owned and managed by owners of the FMCG/supermarket shelves and those owned and managed by manufacturers. With the increasing power and dominance of retailers, it was argued in the literature review that despite the expectations and actions of consumers and manufacturers, retailers have the capacity to significantly influence the final composition of the private label and manufacturer brands offered in the supermarket product categories. This was expected to be especially so in environments characterised by very high retail consolidation and concentration, where the retailers would be expected to have the capacity to employ coercive power to achieve certain outcomes.

This book has examined strategic policy areas in the coexistence of private label and manufacturer brands in the categories, covering the following: private label growth and equilibrium, shelf/merchandising decisions, category management arrangements, driving category growth through innovation and category support, and private label quality spectrum. The study used the bases of power (Hunt 2015; French and Raven 1959) as an interpretive framework. Despite the high retail consolidation and concentration in the New Zealand grocery retail industry and the power imbalance that is in favour of the grocery retail chains in relation to manufacturers/suppliers, and contrary to what one might expect, coercive power was not found to be the dominant source of power governing the coexistence of the two types of brands. As discussed earlier, the coexistence was found to be driven primarily by expert power, and secondarily by referent power. Reward power is also at play, and all these bases of power are non-coercive. Therefore, the relevance of research propositions 7a and 7b is supported.

This book argues that the outcome with respect to the use of power is partly a reflection of the strategic dependency between private label and manufacturer brands in the categories. Dapiran and Hogarth-Scott (2003) suggested that, where there is high retail concentration and low grocery retailer dependence on the supplier, retailers are more likely to employ coercive power. Logically, therefore, it can be reasoned further that the strategic dependency between private label and manufacturer brands in the areas of innovation and category support is high. Additionally, this state of affairs can be further understood from the perspective that, in the category management relationship between manufacturers and retailers, each party brings something that is valued by the other party to the table. Therefore, despite the fact that it is widely recognised in the literature that power in the FMCG sector has largely shifted from manufacturers to retailers and that the balance of power is in the hands of the retailers, there is still an intricate power–dependence relationship that plays a role in shaping the nature of the coexistence between the two types of brands. The concept of countervailing power (Hunt and Nevin 1974; Howe 1990) is very much at play in the categories. Additionally, the state of affairs is also a reflection of retailer strategic thinking on how the power issue should be navigated, especially taking into account the long-term strategic health of the product categories.

Resultant Modified Conceptual Framework

The resultant modified conceptual framework is shown in Fig. 9.1. As can be seen in the framework, the categories that were studied have been incorporated. This is partly in recognition of the delimitations of the scope of the study with respect to analytic generalisation. Furthermore, in connection with the other concepts in the framework, product innovation and category support, alongside consumer choice, were confirmed in the study as having relevance to the determination of strategic management regimes governing the coexistence of private label and manufacturer brands in FMCG/supermarket product categories. Product innovation and category support are in fact areas of strategic dependency. Mainstream academic literature has not spelt out the importance of product innovation and category support in this area, although consumer choice has been a commonly recognised variable.

Additionally, an assessment of the role played by the bases of power in terms of which bases are dominant (particularly in a highly consolidated and concentrated FMCG/supermarket environment) shows that expert and referent bases of power, rather than coercive power, are dominant. It can be argued that this is a reflection of retailer strategic thinking, as the results of the research have shown that retailers are very much in charge and dictate matters as far as the coexistence of the two types of brands in the categories is concerned.

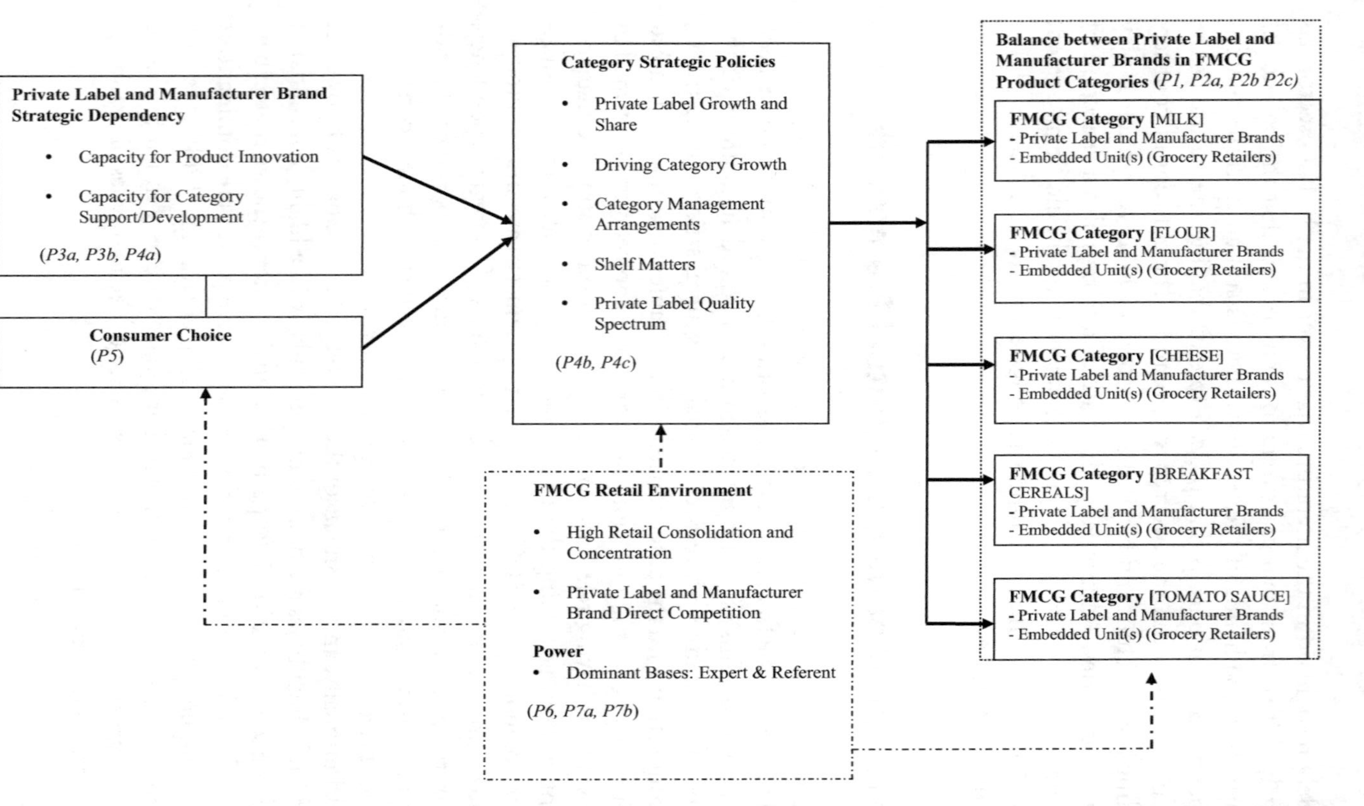

Fig. 9.1 Private label and manufacturer brands in FMCG product categories: modified conceptual framework and research propositions

Theoretical Implications

The theoretical implications of this study are presented in this section in the context of the answers to the research questions raised by the book and the contribution of the research to the academic literature. The primary research question, which was derived through a literature review, concerns how private label and manufacturer brands coexist in FMCG/supermarket product categories in an FMCG landscape characterised by high retail concentration, and how relevant power is to this coexistence. The primary research question was further divided into three subsidiary research questions. Answers to the subsidiary research questions provide an answer to the primary research question, and these answers form the bulk of the academic contribution of the research.

The first subsidiary research question was: Does a grocery retail environment characterised by high retail concentration lead to an overdominance of private label in relation to manufacturer brands in FMCG/supermarket product categories? Based on the results of the study, the answer to this question is that such a grocery retail environment does not necessarily lead to overdominance of private label brands over manufacturer brands in FMCG/supermarket product categories (Chimhundu et al. 2011). There are equilibrium points in the coexistence of the two types of brands in the categories, beyond which grocery retailers may not want to take their private label, and these equilibrium points safeguard the long-term strategic health of the categories.

A review of the literature established the trend of increased grocery retail consolidation and concentration, especially in some economies, and showed that grocery retail consolidation/concentration fuels private label brand shares to high levels, thereby giving the impression that highly consolidated grocery retail environments are synonymous with high private label penetration. Given the increasing power of retailers, and the retailers' ambitions with respect to their private labels, it was necessary to establish whether this might fuel private label growth and share with no end in sight, especially given that the literature has shown that private label brands have a higher capacity to generate profit. It was important for an investigation of this issue to include a highly concentrated FMCG/supermarket contextual environment, and New Zealand is one such environment.

The second subsidiary research question was: How important are aspects of strategic dependency between manufacturer brands and private label in determining the nature of the coexistence between the two types of brands in FMCG/supermarket product categories? The answer to this question is that aspects of strategic dependency between private label and manufacturer brands have relevance for the determination of policies that govern the coexistence of the two types of brands in the product categories. These aspects of strategic dependency are in the areas of comparative capacity to deliver product innovation, and category support on the part of private label and manufacturer brands. The greater collective capacity of manufacturer brands to deliver product innovation and category support is an important aspect of strategic dependency that helps to shape the nature of coexistence between the two types of brands. Therefore, while consumer choice is prominent in the academic literature (through its central role in category management) as a factor that plays a part in the determination of the nature of coexistence between private label and manufacturer brands in the categories, this research has shown that other deeper, underlying factors that have not been given much prominence in the mainstream academic literature are at play as well.

In addition, in relation to the areas of strategic dependency incorporated into this study, the FMCG marketing literature has largely portrayed manufacturer brand innovation (and related marketing activities) as a competitive tool that is employed against private label, in addition to competing with other manufacturer brands, and suggested that this inhibits private label in the act of competing with them. The alternative view, of manufacturer brand innovation having a positive impact on private label, has hardly been investigated in depth. This was worth examination, especially given that the imitative and parasitic behaviour associated with private label is indirect testimony to the fact that private label brands have something to gain from the activities of manufacturer brands. The study has shown that aspects of manufacturer brand innovation and brand/category support that inhibit private label seem to be outweighed by those that enhance private label. The competitiveness of innovative manufacturer brands does inhibit private label; yet private label is enhanced by manufacturer brand innovation driving the categories and category growth. Moreover, the adaptation of successful manufacturer brand innovations by private label enhances private label, and so does the customer pulling power of innovative manufacturer brands.

These are areas in which private label is strategically dependent on manufacturer brands. Thus, manufacturer brand innovation and category support largely have a positive impact on the welfare of private label in FMCG/supermarket categories. What this book proposes is that manufacturer brand innovation and marketing are largely positive and welcome to the private label. They help to shape retailer strategic policies on the coexistence of the two types of brands in the categories. The logical conclusion from this is therefore that more innovative and supportive manufacturer brands are good news for private label. This offers a fresh perspective in explaining the coexistence of private label and manufacturer brands in grocery retail categories, especially given that the two are in a state of direct competition on the shelves. Aspects of manufacturer brand innovation and category support that enhance private label play a strategic role.

The third subsidiary research question was: In an FMCG/supermarket landscape characterised by high retail concentration and direct competition between brands owned and managed by owners of the grocery retail shelves (private label) and those owned and managed by manufacturers (manufacturer brands), what role is played by power in the coexistence relationship between the two types of brands in FMCG/supermarket product categories? The answer to this question, based on the New Zealand study, is that in such a landscape, as well as in the situation of direct competition between private label and manufacturer brands, the theory of power is relevant to the coexistence of the two types of brands in an intricate manner. Despite the balance of power being very much in favour of grocery retailers, it is largely expert and referent bases of power rather than coercive power that are dominant in the coexistence relationship of the two types of brands.

By implication, as an overall assessment of the research contribution of the book, the above contributions are taken to be examples of the "smaller bricks of new knowledge" (Lindgreen et al. 2001: 513) that advance the academic literature. They add an "incremental step in understanding" (Phillips and Pugh 2000: 64) the coexistence of private label and manufacturer brands in FMCG/supermarket product categories in the FMCG landscape under discussion.

Additionally, this book on the coexistence of private label and manufacturer brands in the product categories has, as illustrated in the literature development discussion, addressed interlinked issues in need of research that advances the academic literature. The literature review showed that these interlinked issues are represented by research directions recommended by a number of authors, and these research directions have been creatively made to systematically converge on the coexistence of private label and manufacturer brands in a way that legitimately advances the literature on the topic.

Furthermore, it was shown in the literature review that most academic studies in the area of this research were carried out in the USA, the UK and Europe. Countries such as New Zealand and Australia have not served as research environments for much of the published academic work. Private label research experts have therefore identified the need to take the research to other environments than those that are frequently researched. Taking the research to New Zealand has made a contribution to the academic literature in a way that offers a fresh perspective, as this under-researched environment offers different conditions in comparison to the USA and Europe. Specifically, the New Zealand FMCG/supermarket industry is much more consolidated than any other in the developed world; it is a duopoly, and its private label portfolio is still in the development stages. The highest private label quality tiers have not been fully utilised.

Once again, from a methodological perspective, it was noted in the literature review table (Table 4.1: Key literature on manufacturer brands and private label) in Chap. 4, that most of the research on private label and manufacturer brands has employed the quantitative methodology. Qualitative empirical studies that allow in-depth analysis of specific issues concerning the coexistence of the two types of brands have not really featured. The research area lacked qualitative, empirical studies, and this research book has largely adopted one.

Implications for Marketing Practice

This book has established that there are equilibrium points in the coexistence of private label and manufacturer brands in FMCG/supermarket categories, thus underlining the importance of both types of brands to the grocery retailers. From the category management perspective, while the optimum balance between private label and manufacturer brands has largely been understood to involve adopting a mix that maximises current category sales and profit, the impact that this has on the long-term strategic health of the categories from an innovation and category support point of view is not certain. The results of this study would seem to suggest that an alternative model that takes into account the long-term strategic health of the categories from an innovation and category support point of view is equally legitimate. Therefore, a category managerial effort that constantly and systematically seeks to establish the appropriate equilibrium between private label and manufacturer brands, and that ensures the maximum level of category innovation, support and growth, would be desirable. This book suggests that these optimum points exist. From a practical perspective, a systematic approach on the part of the retail chains to establish where the points are in each category would be a step forward. It should be noted, though, that there would not be a standard formula for this because category situations differ. So a full account of the circumstances of each category would have to be taken into consideration.

The book has also established that aspects of manufacturer brand innovation and category support that enhance private label are taken into account in the determination of category-specific strategic management regimes relating to the coexistence of the two types of brands in the categories. Despite promoting the competitiveness of manufacturer brands, the innovation and marketing activities of manufacturer brands are good for private label. Manufacturers/suppliers have to take note of the fact that, in the new grocery retail environment characterised by high retail consolidation and concentration, as well as retailer power, manufacturer brand innovation and marketing are key expert power sources that can shape the nature of the coexistence of their brands with private label in the categories. Manufacturer brands that excel in this respect are viewed in a more favourable light by the powerful retail chains than those that do not. In addition, the study has shown that the referent power of strong manufacturer brands is desirable to the private label. Therefore, investment in brand equity through product innovation and marketing support is a sure way for the FMCG manufacturer brand to achieve continued coexistence with the private label in the product categories, especially in this era of category rationalisation.

Global Implications and Comparative Case Study Examples

The findings of this book have global implications beyond New Zealand, as has already been discussed, and this can be further illustrated by making comparisons between New Zealand and other countries using a standard set of theoretical points. To do so, comparisons are made between four consumer goods industries with private label brands. The focus of this comparison is the interplay of two key factors, retail consolidation and retailer private label strategy, and how this has practical implications for private label growth in relation to manufacturer brands. For the sake of this exercise, low private label penetration is considered to be anything below 15% private label share in relation to manufacturer brands; medium private label penetration is anything from 15% to 29%, and high private label penetration is anything from 30% upwards.

A Case Study of Switzerland and the UK

The food/grocery retail industries of these two countries have been selected for having a very high private label share of approximately 45% (Switzerland) and approximately 41% (the UK). Retail consolidation is measured by the use of a concentration ratio, which is the percentage of sales commanded by the largest retail company in an industry (Defra 2006). A five-firm concentration ratio of 60% and above is considered high. Both Switzerland and the UK are highly consolidated and have high retail concentration ratios. Coming on to private label portfolio, both Switzerland and the UK employ the full private label quality spectrum of first-, second-, third- and fourth-generation private labels. Therefore, they have premium private label brands that match premium manufacturer brands in their private label portfolio. Thus high private label penetration was achievable.

A Case Study of Australia

The food/grocery retail industry of this country has been selected for having a medium private label share of approximately 21%. Australia is also highly consolidated and has a high retail concentration ratio. As for private label portfolio, Australia has employed first-, second- and third-generation brands, and fourth-generation private label is arguably well in place as well. This may be one explanation, among others, for the steady growth in private label.

A Case Study of New Zealand

The food/grocery retail industry of this country has been selected for having a low private label share of approximately 13%. Likewise, New Zealand is also highly consolidated and has a very high retail concentration ratio. The New Zealand situation is slightly different in that, with respect to private label portfolio, first-, second- and third-generation private labels feature. The fourth generation could still be something of the future, but some marketing practitioners argue that the market is too small to support the development of premium private labels and benefit from that. This may partly explain the stagnation of private label share in this industry.

Implications for Private Label Growth

Retail consolidation in the consumer goods industry results in large scale FMCG retail organisations. The resultant scale generally gives private label "product innovation, consumer research and marketing muscle" (Nielsen 2014: 22). An analysis of retail concentration (a result of retail consolidation) and the employment of a private label quality spectrum can be done for any private label industry globally, and depending on industry actions and private label strategy concerning the employment of either a partial or a full private label quality spectrum, business practitioners and scholars will be better able to comprehend and predict private label trends in any market on the globe.

Drawing Conclusions from this Book

This book has employed the multiple-case study methodology. The methodological limitations of this approach were discussed in Chap. 7, and it is important to note that the book has addressed the potential weaknesses of this methodology by adopting an approach that is considered by case research methodology experts (e.g. Yin 2003, 2013) to be rigorous and systematic. The approach has included making use of prior theory, using multiple sources of evidence, establishing a chain of evidence, allowing interview participants to review interview data (and category observation data where applicable), addressing rival explanations, using replication logic, using a case study protocol and developing a case study database. Additionally, systematic analysis of case study evidence was largely based on the recommendations of relevant research experts (e.g. Gibbs 2007; Miles and Huberman 1994). Therefore, the research boasts a good measure of validity and reliability.

When drawing conclusions from this work, it should be borne in mind that the book seeks analytic generalisation. The research issues identified have therefore been tested for adequacy in the context of the delimitations of the scope of the research. While its scope is limited to the categories studied, the grocery retail chains and the FMCG/supermarket industry and economy studied, the book has a global appeal from the perspective that lessons learnt from this book would likely hold true for industries in other parts of the world with similar conditions. Furthermore, the lessons should be of interest to all grocery industry environments that have both private label and manufacturer brands on grocery retail shelves.

Directions for Further Research

This volume offers opportunities for further research from a number of perspectives. Firstly, the study largely focused on five food product categories that have a private label brand presence: milk, flour, cheese, breakfast cereals and tomato sauce. Future research could investigate other FMCG/supermarket categories to see if the same findings hold true there as well. Secondly, the main study was based on the New Zealand FMCG/supermarket sector. Industry environments would usually differ in one or more areas from economy to economy. The research could be extended to FMCG industries in other countries. Such studies could still maintain the case study approach, which allows an in-depth investigation of the specific FMCG industries chosen. Thirdly, it was noted that the nature of the study undertaken was such that the findings of the research are generalisable to theory (i.e. analytic generalisation). The study could be further extended to achieve statistical generalisation. This would mean designing research that largely adopts the positivist approach, covering many categories and a number of economies. In this regard, the modified conceptual framework created in this book has research propositions that have been advanced, and that can be refined into hypotheses that could then be empirically tested using studies set in the positivist paradigm. Fourthly, a comparative study of a number of FMCG industries that have a high level of retail concentration and those that have a low level of retail concentration could be conducted to establish whether there are fundamental differences between the two groups with respect to retailer strategic thinking and general approach to the coexistence of the two types of brands. Fifthly, the private label phenomenon is not only limited to FMCG; it does exist in other sectors as well. Future studies could investigate the coexistence of private label and manufacturer/supplier brands in industries such as business-to-business and services.

Chapter Recap

This final chapter of this book has addressed a summary statement of outcomes on the research issues, findings of the research in the context of the literature, resultant modified conceptual framework, theoretical implications and implications for marketing practice, drawing conclusions from the study and suggesting directions for further research.

Conclusion

This book has investigated the coexistence of private label (i.e. retailer own brands) and manufacturer brands in FMCG/supermarket product categories in an environment characterised by high retail consolidation and concentration, as well as direct competition between brands owned and managed by owners of the grocery retail shelves (private label) and those owned and managed by their suppliers (manufacturer brands). The book is set in the interpretive paradigm (realist variant) and has adopted the multiple-case study research methodology.

The book has argued firstly that a grocery retail environment characterised by high retail consolidation and concentration does not necessarily lead to an overdominance of private label in relation to manufacturer brands in FMCG/supermarket product categories (Chimhundu et al. 2011). Secondly, the book has reasoned that aspects of strategic dependency between private label and manufacturer brands have relevance to the determination of policies that govern the coexistence of the two types of brands in FMCG/supermarket product categories (Chimhundu et al. 2010). Furthermore, manufacturer brands' greater collective capacity to deliver product innovation and category support is a key aspect of the strategic dependency between private label and manufacturer brands that shapes the nature of coexistence between the two types of brands (Chimhundu et al. 2015a, b). Finally, the book has further reasoned that in an FMCG landscape characterised by high retail consolidation and concentration, and direct competition between brands owned and managed by owners of the grocery retail shelves and those owned and managed by their suppliers, power is relevant to the coexistence of the two types of brands in an intricate manner. Although the balance of power is in favour of the grocery retailers, it is largely expert and referent bases of power rather than coercive power that dictate the coexistence relationship of the two types of brands (Chimhundu 2016).

The book has further provided a resultant modified conceptual framework that systematically integrates the outcomes of the key research issues investigated, and this framework comes with research propositions that individually offer opportunities and directions for further research. These research propositions could be further developed into hypotheses that could be investigated using confirmatory studies that are set in the positivist paradigm. Therefore, in addition to this book's contribution to knowledge and practice, it also offers a wide variety of avenues for further intellectual enquiry that would contribute to knowledge and shape practice in the area of marketing private label and manufacturer brands. Finally, and most importantly, the underlying lessons from this book are relevant globally and should hold true in other environments around the world that may have more or less similar conditions.

References

ACNielsen. (2005). *The power of private label: A review of growth trends around the world*. New York, NY: ACNielsen.

ACNielsen, Karolefski, J., & Heller, A. (2006). *Consumer-centric category management: How to increase profits by managing categories based on consumer needs*. Hoboken, NJ: Wiley.

Anonymous. (2004). Effective innovation. *Strategic Direction, 20*(7), 33–35.

Anonymous. (2005). House brand strategy doesn't quite check out. *The Age*. Retrieved December 3, 2017, from http://www.theage.com.au/news/Business/House-brand-strategy-doesnt-quite-check-out/2005/04/01/1112302232004.html

Aribarg, A., Arora, N., Henderson, T., & Kim, Y. (2014). Private label imitation of a national brand: Implications for consumer choice and law. *Journal of Marketing Research, 51*(6), 657–675.

BCG (Boston Consulting Group). (2005). *Innovation 2005*. Boston, MA: The Boston Consulting Group Inc.

Berthon, P., Hulbert, J. M., & Pitt, L. F. (1997). Brands, brand managers and the management of brands: Where to next? *MSI Report No. 97–122*. Cambridge, MA: Marketing Science Institute.

Booz, Allen and Hamilton. (1982). *New product management for the 1980s*. New York, NY: New York Press.

Brenner, M. S. (1994). Tracking new products: A practitioner's guide. *Research Technology Management, 37*(6), 36–40.

Burt, S. (2000). The strategic role of retail brands in British grocery retailing. *European Journal of Marketing, 34*(8), 875–890.

Burt, S. L., & Sparks, L. (2003). Power and competition in the UK retail grocery market. *British Journal of Management, 14*(3), 237–254.

Chimhundu, R. (2016). Marketing store brands and manufacturer brands: Role of referent and expert power in merchandising decisions. *Journal of Brand Management, 23*(5), 24–40.

Chimhundu, R., Hamlin, R. P., & McNeill, L. (2010). Impact of manufacturer brand innovation on retailer brands. *International Journal of Business and Management, 5*(9), 10–18.

Chimhundu, R., Hamlin, R. P., & McNeill, L. (2011). Retailer brand share statistics in four developed economies from 1992 to 2005: Some observations and implications. *British Food Journal, 113*(3), 391–403.

Chimhundu, R., Kong, E., & Gururajan, R. (2015a). Category captain arrangements in grocery retail marketing. *Asia Pacific Journal of Marketing and Logistics, 27*(3), 368–384.

Chimhundu, R., McNeill, L. S., & Hamlin, R. P. (2015b). Manufacturer and retailer brands: Is strategic coexistence the norm? *Australasian Marketing Journal, 23*(1), 49–60.

Coelho do Vale, R., & Verga-Matos, P. (2015). The impact of copycat packaging strategies on the adoption of private labels. *Journal of Product & Brand Management, 24*(6), 646–659.

Collins-Dodd, C., & Zaichkowsky, J. L. (1999). National brand responses to brand imitation: Retailers versus other manufacturers. *Journal of Product and Brand Management, 8*(2), 96–105.

Conn, C. (2005). Innovation in private label branding. *Design Management Review, 16*(2), 55–62.

Coriolis Research. (2002). *Responding to private label in New Zealand*. Auckland: Coriolis Research.

Cotterill, R. W. (1997). The food distribution system of the future: Convergence towards the US or UK model? *Agribusiness, 3*(2), 123–135.

Dapiran, G. P., & Hogarth-Scott, S. (2003). Are co-operation and trust being confused with power? An analysis of food retailing in Australia and New Zealand. *International Journal of Retail and Distribution Management, 31*(5), 256–267.

Defra (Department for Environment, Food and Rural Affairs). (2006). *Economic note on UK grocery retailing*. London, UK: Department for Environment, Food and Rural Affairs, Food and Drinks Economics Branch.

Doyle, P., & Bridgewater, S. (1998). *Innovation in marketing*. Oxford: Butterworth-Heinemann.

French, J. R. P., & Raven, B. (1959). The bases of social power. In D. Cartwright (Ed.), *Studies in social power*. Institute of Social Research (pp. 150–167), Ann Arbor, MI: The University of Michigan.

Galbraith, J. K. (1952). *American capitalism: The concept of countervailing power*. Boston, MA: Houghton Mifflin Company.

Gibbs, G. R. (2007). *Analyzing qualitative data*. London: Sage.

Guinet, J., & Pilat, D. (1999). Promoting innovation: Does it matter? *The OECD Observer*, No. 217/218, Paris, pp. 63–65.

Hardaker, G. (1998). An integrated approach towards product innovation in international manufacturing organizations. *European Journal of Innovation Management, 1*(2), 67–73.

Harvey, M. (2000). Innovation and competition in UK supermarkets. *Supply Chain Management: An International Journal, 5*(1), 15–21.

Hoch, S. J., & Banerji, S. (1993). When do private labels succeed? *Sloan Management Review, 34*(4), 57–67.

Hoch, S. J., Montgomery, A. L., & Park, Y. H. (2002a). Why private labels show long-term market evolution. *Marketing Department Working Paper*, Wharton School, University of Pennsylvania, PA.

Hoch, S. J., Montgomery, A. L., & Park, Y. H. (2002b). Long-term growth trends in private label market shares. *Marketing Department Working Paper #00-010*, Wharton School, University of Pennsylvania, PA.

Hogarth-Scott, S. (1999). Retailer-supplier partnerships: Hostages to fortune or the way forward for the millennium? *British Food Journal, 101*(9), 668–682.

Hollingsworth, A. (2004). Increasing retail concentration: Evidence from the UK food sector. *British Food Journal, 106*(8), 629–638.

Hovhannisyan, V., & Bozic, M. (2013). A benefit-function approach to studying market power: An application to the US yogurt market. *Journal of Agricultural and Resource Economics, 38*, 159–173.

Hovhannisyan, V., & Bozic, M. (2016). The effects of retail concentration on retail dairy product prices in the United States. *Journal of Dairy Science, 99*(6), 4928–4938.

Howe, W. S. (1990). UK retailer vertical power, market competition and consumer welfare. *International Journal of Retail and Distribution Management, 18*(2), 16–25.

Hunt, S. D. (2015). The bases of power approach to channel relationships: Has marketing's scholarship been misguided? *Journal of Marketing Management, 31*(7–8), 747–764.

Hunt, S. D., & Nevin, V. R. (1974). Power in a channel of distribution: Sources and consequences. *Journal of Marketing Research, 11*(2), 186–193.

Information Resources, Inc. (2005), Private label: The battle for value-oriented shoppers intensifies. *Times and Trends: A Snapshot of Trends Shaping the CPG Industry*, November, pp. 1–23.

Kaswengi, J., & Diallo, M. F. (2015). Consumer choice of store brands across store formats: A panel data analysis under crisis periods. *Journal of Retailing and Consumer Services, 23*, 70–76.

Kumar, N., & Steenkamp, J. E. M. (2007a). *Private label strategy: How to meet the store brand challenge*. Boston, MA: Harvard Business School Press.

Kumar, N., & Steenkamp, J. E. M. (2007b). Brand versus brand. *International Commerce Review, 7*(1), 47–53.

Kung, H., & Schmid, L. (2015). Innovation, growth, and asset prices. *The Journal of Finance, 70*(3), 1001–1037.

Kurtulus, M., & Toktay, L. B. (2005). Category captaincy: Who wins, who loses? *ECR Journal, 5*(1), 59–65.

Laaksonen, H. (1994). *Own brands in food retailing across Europe*. Oxford: Oxford Institute of Retail Management.

Laaksonen, H., & Reynolds, J. (1994). Own brands in food retailing across Europe. *The Journal of Brand Management, 2*(1), 37–46.

Lehmann, D. R., & Winer, R. S. (2002). *Product management* (3rd ed.). New York, NY: McGraw-Hill.

Lindgreen, A., Vallaster, C., & Vanhamme, J. (2001). Reflections on the PhD process: The experience of three survivors. *The Marketing Review, 1*(4), 505–529.

Lindsay, M. (2004). Editorial: Achieving profitable growth through more effective new product launches. *Journal of Brand Management, 12*(1), 4–10.

Major, M., & McTaggart, J. (2005, November 15). Blueprints for change. *Progressive Grocer*, pp. 89–94.

Miles, M. B., & Huberman, A. M. (1994). *Qualitative data analysis: An expanded sourcebook*. Thousand Oaks, CA: Sage Publications.

Nelson, W. (2002). Practice papers all power to the consumer? Complexity and choice in consumers' lives. *Journal of Consumer Behaviour, 2*(2), 185–195.

Nielsen. (2014). *The state of private label around the world*. The Nielsen Company.

Ogbonna, E., & Wilkinson, B. (1998). Power relations in the UK grocery supply chain: Developments in the 1990s. *Journal of Retailing and Consumer Services, 5*(2), 77–86.

Olbrich, R., Hundt, M., & Jansen, H. C. (2016). Proliferation of private labels in food retailing: A literature overview. *International Journal of Marketing Studies, 8*(8), 63–76.

Panigyrakis, G., & Veloutsou, C. A. (2000). Problems and future of the brand management structure in the fast moving consumer goods industry: The viewpoint of brand managers in Greece. *Journal of Marketing Management, 16*(1/3), 165–184.

Phillips, E. M., & Pugh, D. S. (2000). *How to get a PhD: Handbook for students and their supervisors* (3rd ed.). Buckingham: Open University Press.

Porter, M. E. (1976). *Interbrand choice, strategy and market power*. Harvard, MA: American Marketing Association.
Putsis, W. P., & Dhar, R. (1996). Category expenditure and promotion: Can private labels expand the pie? *Working Paper*, Yale University, New Haven, CT.
Rizkallah, E. G., & Miller, H. (2015). National versus private-label brands: Dynamics, conceptual framework, and empirical perspective. *Journal of Business & Economics Research, 13*(2), 123–136.
Robert, M. (1995). *Product innovation strategy: Pure & simple*. New York, NY: McGraw-Hill.
Silverman, D. (2004). *Interpreting qualitative data: Methods for analysing talk, text and interaction* (2nd ed.). London: Sage Publications.
Sirimanne, E. (2016). Private label: Global, New Zealand & Australian perspectives. *Euromonitor International*. Retrieved September 16, 2017, from www.blog.euromonitor.com
Stanković, L., & Končar, J. (2014). Effects of development and increasing power of retail chains on the position of consumers in marketing channels. *Ekonomika Preduzeća, 62*(5–6), 305–314.
Steenkamp, J. B. E., & Dekimpe, M. G. (1997). The increasing power of store brands: Building loyalty and market share. *Long Range Planning, 30*(6), 917–930.
Steenkamp, J. B. E., Van Heerde, H. J., & Geyskens, I. (2010). What makes consumers willing to pay a price premium for national brands over private labels? *Journal of Marketing Research, 47*(6), 1011–1024.
Steiner, R. L. (2004). The nature and benefits of national brand/private label competition. *Review of Industrial Organization, 24*(2), 105–127.
Sutton-Brady, C., Kamvounias, P., & Taylor, T. (2015). A model of supplier–retailer power asymmetry in the Australian retail industry. *Industrial Marketing Management, 51*, 122–130.
Verhoef, P. C., Nijssen, E. J., & Sloot, L. M. (2002). Strategic reactions of national brand manufacturers towards private labels: An empirical study in The Netherlands. *European Journal of Marketing, 36*(11/12), 1309–1326.
Weitz, B., & Wang, Q. (2004). Vertical relationships in distribution channels: A marketing perspective. *Antitrust Bulletin, 49*(4), 859–876.
Yin, R. K. (2003). *Case study research: Design and methods* (3rd ed.). Applied Social Research Methods Series, Vol. 5. Thousand Oaks, CA: Sage Publications.
Yin, R. K. (2013). *Case study research: Design and methods* (5th ed.). Thousand Oaks, CA: Sage Publications.

Appendix

List of Codes Used in Data Management (Research Interview Data)

BAL: Balance (between manufacturer brands and private label in FMCG/ supermarket product categories)

BAL-SHA: Balance relating to share
BAL-SS: Balance relating to shelf space
BAL-SF: Balance relating to shelf facings
BAL-SH/P: Balance relating to shelf height/position
BAL-NOB: Balance on number of brands
BAL-NOP: Balance on number of products
BAL-RET-CHN: Balance in the retail chain

EQ: Equilibrium

RET-CONC: Retail concentration
RET-CONS: Retail consolidation

R. Chimhundu, *Marketing Food Brands*,
https://doi.org/10.1007/978-3-319-75832-9

SUPP-CONC: Supplier concentration
SUPP-CONS: Supplier consolidation

CONS-PERCE: Consumer perceptions
CONS-CH: Consumer choice

- CONS-CH-VA: Consumer choice from a variety perspective
- CONS-CH-POC: Consumer choice from a power of choice perspective

VA: Variety

- VA-NOB: Variety relating to number brands
- VA-NOP: Variety relating to number of products

STRA-DEP: Strategic dependency
ENH-RB: Enhancing retailer brand
INH-RB: Inhibiting retailer brand
RET-PHIL: Retailer philosophy/RET-STRA: Retailer strategy
INT-DEP: Interdependence

CAP-IN: Capacity for innovation

MB-CAP-IN: Manufacturer brand capacity for innovation
RB-CAP-IN: Retailer brand capacity for innovation
RE-IN: Resources for innovation
MB-RE-IN: Manufacturer brand resources for innovation
RB-RE-IN: Retailer brand resources for innovation
EX-IN: Expertise in innovation
MB-EX-IN: Manufacturer brand expertise in innovation
RB-EX-IN: Retailer brand expertise in innovation
RO-IN: Rate of innovation
MB-RO-IN: Manufacturer brand rate of innovation
RB-RO-IN: Retailer brand rate of innovation
INC-IN: Incremental innovation

- RA-IN: Radical innovation
- QUAL-IN: Quality of innovation

- INCE-IN: Incentives for innovation
- DINCE-IN: Disincentives for innovation
- IN-PATT: Innovation patterns
- BACK-INTEG: Backward integration

CAP-CDEV: Capacity for category development

MB-CAP-CDEV: Manufacturer brand capacity for category development
RB-CAP-CDEV: Retailer brand capacity for category development
RE-CDEV: Resources for category development

- MB-RE-CDEV: Manufacturer brand resources for category development
- RB-RE-CDEV: Retailer brand resources for category development

EX-CDEV: Expertise in category development

- MB-EX-CDEV: Manufacturer brand expertise in category development
- RB-EX-CDEV: Retailer brand expertise in category development

INCE-CDEV: Incentives for category development

- DINCE-CDEV: Disincentives for category development

RET-AW-IN: Retailer awareness of manufacturer brand superior capacity for innovation
RET-AW-CDEV: Retailer awareness of manufacturer brand superior capacity for category development

BRA: Branding/brand strategy

BRA-TRU: Brand trust
NICHE: Niche strategy
DIFF: Differentiation
PR-DR: Price-driven
QUAL-MB/RB: Quality of manufacturer brand or retailer brand
PREM-MID-BU: Premium, middle and budget segments
PA: Patents
LE-SU: Legal suit(s)

MAR: Margin(s) (PRO: Profit)
RET-BR-ST: Retailer brand strategy
MB-ST: Manufacturer brand strategy
POL: Political aspects of retailer brand or manufacturer brand marketing
CUS-PUL: Customer pull

PO: Power

EXP-PO: Expert power
COE-PO: Coercive power
REW-PO: Reward power
REF-PO: Referent power
LE-PO: Legitimate power

STRA-REG: Strategic management regimes

STRA-REG-RB-GR: Strategic management regimes on retailer brand growth/limit/share
STRA-REG-IN: Strategic management regimes on brand/product innovation
STRA-REG-CDEV: Strategic management regimes on category development
STRA-REG-CMAR: Strategic management regimes on category management arrangements
STRA-REG-SHLF: Strategic management regimes on shelf matters (merchandising) (stocking (STO); shelf space (SS), shelf facings (SF), shelf height/position (SH/P); product deletion (PDEL); rationalisation (RAT); level of consumer choice (CONS-CH))
STRA-REG-RB-QUAL: Strategic management regimes on retailer brand quality spectrum
STRA-REG-HLTH: Strategic management regimes on category strategic health
STRA-REG-DET: Strategic management regime determination
STRA-CONT: Strategic control

NZ-ECO: New Zealand economic situation

RE-STRA-OB: Retailer strategic objectives
SM-NZ-MKT: Small nature of the New Zealand market
MB-CON-IN&CDEV: Manufacturer brand contributions to innovation and category development
RB-CON-IN&CDEV: Retailer brand contributions to innovation and category development
DEP-CAP-IN: Dependency on capacity for innovation
DEP-CAP-CDEV: Dependency on capacity for category development
TA: Tacit understanding
STO-AUTO: Store autonomy

CO-BA: Company background
CATEG-CHA: Category characteristics
TECH: Technology (TECH-SOPH: Technological sophistication)
CATEG-SUPP: Category supply
PRODN-CAPA: Production capacity
CAT: Categories (MIL, milk; FL, flour; CHE, cheese; BR, breakfast cereals; TS, tomato sauce)
COMPE: Competition
DEMOG: Demographics
WA-DO: Watchdog
MKT-FO: Market forces

Glossary

Brand management The coordination of marketing activities for a specific brand (product) that includes the development and implementation of the brand marketing plan and the monitoring of performance of the brand[1].

Category management A joint, manufacturer (supplier)–retailer process of defining and managing product categories as strategic business units, focusing on satisfying consumer needs, and with the objective of producing enhanced business results.

Category strategic management regimes Strategic policies governing the FMCG/supermarket product categories.

Category support (or category marketing support) Any marketing activities other than product innovation and its commercialisation that help to develop and grow the FMCG/supermarket categories (e.g. advertising, sales promotion, merchandising support, monies/revenues paid to the retailers by suppliers, market research, branding/brand development and brand management). This can also be referred to as category development.

Consumer choice The range of products stocked on the shelves for consumers to choose from (or the provision of such a range), as well as the consumers' liberty to make their choice. In this respect, competition amongst the products is considered to be a reflection of the respective consumer choice.

[1] These definitions are given as employed in the book. Some of the definitions are developed for this book in the literature chapter (Chap. 2).

R. Chimhundu, *Marketing Food Brands*,
https://doi.org/10.1007/978-3-319-75832-9

The author/the researcher The writer of this book; the researcher of the large project that forms the basis of this book.

Research participants Individuals (or organisations) interviewed by the author during the data collection process and/or who supplied data for the book.

Private label (or private label brands) Brands that are owned and managed by food and grocery retail chains. These brands can also be referred to as retailer own brands, own-label goods, store brands, house brands, private brands, distributor brands, retail brands, retailer brands, home brands, generic brands or generic products.

Manufacturer brands Brands that are owned and managed by manufacturers. These can also be referred to as national brands (and they can be regional, national or international/global brands).

Fast-moving consumer goods Branded and packaged products that are largely sold through supermarkets. These are also referred to, in some parts of the world, as consumer packaged goods. They consist of food and grocery products.

Grocery retail chain A retail company that has supermarket groups in its business portfolio.

Product category A group of products that have similar characteristics and that satisfy similar end-user needs.

Product innovation The creation of new or modified (i.e. updated) products/brands. The scope ranges from completely new, breakthrough concepts to minor adjustments (incremental changes). New product development is a key facet of product innovation. Related marketing activities that come with the innovation are taken as part and parcel of the innovation.

Retail consolidation/retail concentration "The concentration of market share in the hands of fewer, larger operators [i.e. retailers]" (IGD.com)

Shelf facing "The physical (linear) space that one product occupies on the shelf or fixture" (IGD.com). Shelf facings are expressed in numerical terms (e.g. 25 facings of such and such a brand or product).

Shelf position The shelf height (or fixture height) or general position at which a product is displayed (e.g. eye-level position, bottom shelf, shelf-end).

Shelf space The width of shelf in a supermarket occupied by a particular brand or product. This can be measured and expressed in centimetres or metres.

Stock keeping units Uniquely identifiable units of a product that comprise different variations of the product (IGD.com).

Store/supermarket/retail site A specific branch of a supermarket chain/group.

Supermarket chain (supermarket group) A group of supermarket stores that fall under the same brand name.

Index

R. Chimhundu, *Marketing Food Brands*,
https://doi.org/10.1007/978-3-319-75832-9

CPSIA information can be obtained
at www.ICGtesting.com
Printed in the USA
LVHW01*1213130518
577031LV00012B/602/P